Synthesis Lectures on Mathematics & Statistics

This series includes titles in applied mathematics and statistics for cross-disciplinary STEM professionals, educators, researchers, and students. The series focuses on new and traditional techniques to develop mathematical knowledge and skills, an understanding of core mathematical reasoning, and the ability to utilize data in specific applications.

Haiyan Tian

Linear Algebra

Key Ideas and Methods for a First Course

Springer

Haiyan Tian
The University of Southern Mississippi
Hattiesburg, MS, USA

ISSN 1938-1743 ISSN 1938-1751 (electronic)
Synthesis Lectures on Mathematics & Statistics
ISBN 978-3-031-84646-5 ISBN 978-3-031-84647-2 (eBook)
https://doi.org/10.1007/978-3-031-84647-2

This Springer imprint is published by the registered company Springer Nature Switzerland AG
The registered company address is: Gewerbestrasse 11, 6330 Cham, Switzerland

To my husband, Jan Laurens, and To my children,
Xianglong and Yolanda

—Haiyan Tian

A Note to the User

This textbook is designed as an introductory one-semester course for STEM majors. As a matter of fact, it can be used by any student who needs to take a foundational mathematics course.

The organization of the materials is the result of my experience of more than 20 years teaching the subject to majors in STEM, including computer science and engineering, ocean engineering, polymer science and engineering, etc. I have decided that it is time to propose a more student-friendly presentation of the material. The key ideas and methods are covered and arranged in an order as intuitive as possible. An advanced background in mathematics is not required from the students. Each topic is followed by a reasonable number of exercises to check on complete understanding of essentials. Clarity, rather than quantity, is a feature the book aims at that will be appreciated by both instructors and their students.

Haiyan Tian

Contents

Systems of Linear Equations 1

1.1 Introduction to Systems of Linear Equations

Linear Equations

In the xy-plane, we can describe a straight line using an equation of the form

$$a_1x + a_2y = b$$

where a_1, a_2, and b are real numbers, and the two coefficients a_1 and a_2 are not both zero. The equation of the form is a linear equation in the variables x and y.

Example 1 (*Linear Equations in Two Variables*) The equations

$$2x - 6y = 5, \quad y = 4x + 1, \text{ and } y = 10$$

are linear. But the equations

$$x^2 + 2y = 1, \quad \cos x - 3y = 5, \text{ and } 6x + \sqrt{y} = 9$$

are not linear.

More generally, a **linear equation** in the n variables $x_1, x_2, \ldots, x_n$ is defined as an equation that can be expressed in the form

$$a_2x_1 + a_2x_2 + \cdots + a_nx_n = b$$

where $a_1, a_2, \ldots, a_n$, and b are real numbers.

H. Tian, *Linear Algebra*, Synthesis Lectures on Mathematics & Statistics,
https://doi.org/10.1007/978-3-031-84647-2_1

Example 2 (*Linear Equations*) The equations

$$2x_1 - 6x_2 = 5 \text{ and } x_1 + 7x_2 - 4x_3 + \frac{1}{2}x_4 = 5$$

are linear. But the equations

$$x_1 + 7x_2 - 4e^{x_3} + \frac{1}{2}x_4 = 5 \text{ and } x_1 + 7\sqrt{x_2} - 4x_3 + \frac{1}{2}x_4 = 5$$

are not linear.

A **solution** of a linear equation $a_1x_1 + a_2x_2 + \cdots + a_nx_n = b$ is a list of n numbers $(t_1, \ t_2, \ \ldots, \ t_n)$ such that the equation is true when we substitute $x_1 = t_1$, $x_2 = t_2$, ..., $x_n = t_n$. For example, $\left(1, \ -\frac{1}{2}\right)$ is a solution of the equation $2x_1 - 6x_2 = 5$. This equation is a true statement when we substitute $x_1 = 1$, and $x_2 = -\frac{1}{2}$. Another example, (5, 0, 0, 0) is a solution to the equation $x_1 + 7x_2 - 4x_3 + \frac{1}{2}x_4 = 5$. This equation is a true statement when we substitute $x_1 = 5$, $x_2 = 0$, $x_3 = 0$, $x_4 = 0$.

Remark 3 The equation $2x - 6y = 5$ is the same as the equation $2x_1 - 6x_2 = 5$. The only difference is the names of the variables. When there are many variables in an equation, the notation x_i, meaning the ith variable, is a convenient way to represent the variables.

A **system of linear equations** is a set of linear equations involving the same variables. The following is a system of m linear equations in n unknowns,

$$\begin{aligned} a_{11}x_1 + a_{12}x_2 + \cdots + a_{1n}x_n &= b_1 \\ a_{21}x_1 + a_{22}x_2 + \cdots + a_{2n}x_n &= b_2 \\ &\vdots \\ a_{m1}x_1 + a_{m2}x_2 + \cdots + a_{mn}x_n &= b_m \end{aligned}$$

where $x_1, x_2, \ldots, x_n$ are the n variables, and the subscripted $a's$ and $b's$ are constants.

Solutions of Linear Systems

In solving a system of linear equations, we hope to find all possible solutions of the equation. The set of all solutions of a linear system is called the **solution set**. First, let's try to find the solution set for a given system of equations in two variables. Our strategy is to reduce a given linear system to a simpler system. The resulted linear system is simpler looking, but is equivalent to the original system. Two linear systems are said to be **equivalent** if they have the same solution set.

Example 4 (*A Linear System with No Solution*) Let's consider a system in two variables,

$$\begin{aligned} x_1 + x_2 &= 0, \\ x_1 + x_2 &= 1. \end{aligned}$$

If we subtract first equation from the second, we can eliminate both variables x_1 and x_2 to obtain an equation $0 = 1$.

$$\begin{aligned} x_1 + x_2 &= 0, \\ 0 &= 1. \end{aligned}$$

The resulted equation $0 = 1$ is false, independent of the values of x_1 and x_2. Hence, the system has **no solution**. The geometric interpretation is that the two straight lines $x_1 + x_2 = 0$ and $x_1 + x_2 = 1$ are parallel, and there is no point that can be on both lines.

Example 5 (*A Linear System with One Solution*) Let's consider another system in two variables,

$$\begin{aligned} x_1 + x_2 &= 0, \\ -x_1 + x_2 &= 6. \end{aligned}$$

Adding the two equations, we can eliminate the x_1 variable and obtain a new equation $2x_2 = 6$, which involves only one variable.

$$\begin{aligned} x_1 + x_2 &= 0, \\ 2x_2 &= 6 \end{aligned}$$

Multiplying both sides of the equation $2x_2 = 6$ by $1/2$, we obtain $x_2 = 3$. Substituting $x_2 = 3$ into either equation of the given system, we get the value of $x_1 = -3$. The system has **one solution** $(-3, 3)$. The geometric interpretation is that the two straight lines $x_1 + x_2 = 0$ and $-x_1 + x_2 = 6$ intersect at the point $(-3, 3)$. In another word, the point $(-3, 3)$ is on both lines.

Example 6 (*A Linear System with Infinitely Many Solutions*) Let's consider another system in two variables,

$$\begin{aligned} x_1 + x_2 &= 3, \\ 2x_1 + 2x_2 &= 6. \end{aligned}$$

If the second equation subtract twice the first equation, the new second equation will be $0 = 0$.

$$\begin{aligned} x_1 + x_2 &= 3, \\ 0 &= 0. \end{aligned}$$

The equation $0 = 0$ is always true, independent of the values of x_1 and x_2. The first equation implies that the x_2 can be any real number t, and the $x_1 = 3 - t$. The solution set is the set of $(3 - t, t)$ with t being any real number. That is, any point on the straight line $x_1 + x_2 = 3$ is a solution of the given system. This system has **infinitely many solutions**. The geometric interpretation is that the two straight lines $x_1 + x_2 = 3$ and $2x_1 + 2x_2 = 6$ are overlapping. Any point on the first line is also on the second line, and vice versa. So every point on the line is a solution of the system, satisfying both equations.

From the above examples of linear systems involving two variables, we see three possibilities: no solution, exactly one solution, and infinitely many solutions. As a matter of fact, we will prove later the same three possibilities hold for any systems of linear equations. A linear system is said to be **consistent** if it has at least one solution; otherwise it is **inconsistent**. Hence, a linear system that has exactly one solution or infinitely many solutions is consistent, and a linear system that has no solution is inconsistent.

Augmented Matrices

For a system of m linear equations in n unknowns,

$$\begin{aligned} a_{11}x_1 + a_{12}x_2 + \cdots + a_{1n}x_n &= b_1 \\ a_{21}x_1 + a_{22}x_2 + \cdots + a_{2n}x_n &= b_2 \\ &\vdots \\ a_{m1}x_1 + a_{m2}x_2 + \cdots + a_{mn}x_n &= b_m \end{aligned}$$

we can use a rectangular array of numbers

$$\begin{bmatrix} a_{11} & a_{12} & \cdots & a_{1n} & b_1 \\ a_{21} & a_{22} & \cdots & a_{2n} & b_2 \\ \vdots & \vdots & & \vdots & \vdots \\ a_{m1} & a_{m2} & \cdots & a_{mn} & b_m \end{bmatrix}$$

to keep track of the coefficients of variables in the equations and the constants on the right hand side of equations. This is called the **augmented matrix** of the system. The matrix

$$\begin{bmatrix} a_{11} & a_{12} & \cdots & a_{1n} \\ a_{21} & a_{22} & \cdots & a_{2n} \\ \vdots & \vdots & & \vdots \\ a_{m1} & a_{m2} & \cdots & a_{mn} \end{bmatrix}$$

that keeps track of only the coefficients of variables is called the **coefficient matrix**.

Example 7 (*Coefficient Matrix and Augmented Matrix*) The linear system

$$\begin{aligned} x_1 + x_2 - x_3 &= 2 \\ x_2 + x_3 &= 3 \\ x_3 &= 1 \end{aligned}$$

has the coefficient matrix $\begin{bmatrix} 1 & 1 & -1 \\ 0 & 1 & 1 \\ 0 & 0 & 1 \end{bmatrix}$ and the augmented matrix $\begin{bmatrix} 1 & 1 & -1 & 2 \\ 0 & 1 & 1 & 3 \\ 0 & 0 & 1 & 1 \end{bmatrix}$. This system looks easy to solve since fewer variables appear in the second and third equations. We notice that its coefficient matrix has a few 0 entries. We determine from the third equation that $x_3 = 1$. Then from the second equation, we get $x_2 = 3 - x_3 = 3 - 1 = 2$. Once we have the values of x_2 and x_3, we find from the first equation that $x_1 = 2 - x_2 + x_3 = 2 - 2 + 1 = 1$. Thus the system has exactly one solution $(1, 2, 1)$. We have used the **backward substitution** to solve the system, which has a **triangular form**.

Solving a Linear System

When solving an arbitrary linear system, we apply three elementary operations to reduce the system. The **three elementary operations** are (1) **(Replacement)** replace an equation by its sum with a multiple of another equation; (2) **(Interchange)** interchange any two equations; (3) **(Scaling)** multiply both sides of an equation by a nonzero real number. Each elementary operation results in a new system that is equivalent to the one before the operation. So the reduced system is equivalent to the original system.

Example 8 (*Solving a Linear System*) The linear system

$$\begin{aligned} x_1 + 2x_2 + x_3 &= 3 \\ 3x_1 - x_2 - 3x_3 &= -1 \\ 2x_1 + 3x_2 + x_3 &= 4 \end{aligned}$$

has the coefficient matrix $\begin{bmatrix} 1 & 2 & 1 \\ 3 & -1 & -3 \\ 2 & 3 & 1 \end{bmatrix}$ and the augmented matrix $\begin{bmatrix} 1 & 2 & 1 & 3 \\ 3 & -1 & -3 & -1 \\ 2 & 3 & 1 & 4 \end{bmatrix}$. Let's try to eliminate the x_1 variable from the second and the third equations. We first replace the second equation by the sum of the second and -3 multiple of the first,

$$
\begin{aligned}
x_1 + 2x_2 + x_3 &= 3 \\
-7x_2 - 6x_3 &= -10 \\
2x_1 + 3x_2 + x_3 &= 4
\end{aligned}
$$

Then we replace the third equation by the sum of the third and -2 multiple of the first,

$$
\begin{aligned}
x_1 + 2x_2 + x_3 &= 3 \\
-7x_2 - 6x_3 &= -10 \\
-x_2 - x_3 &= -2
\end{aligned}
$$

We notice the corresponding augmented matrix now as

$$
\begin{bmatrix} 1 & 2 & 1 & 3 \\ 0 & -7 & -6 & -10 \\ 0 & -1 & -1 & -2 \end{bmatrix}
$$

after the replacement of the second row and the third row of the original augmented matrix. For the current system, we interchange the second and third equations to obtain another equivalent system,

$$
\begin{aligned}
x_1 + 2x_2 + x_3 &= 3 \\
-x_2 - x_3 &= -2 \\
-7x_2 - 6x_3 &= -10
\end{aligned}
$$

We can scale the second equation by multiplying its both sides by -1,

$$
\begin{aligned}
x_1 + 2x_2 + x_3 &= 3 \\
x_2 + x_3 &= 2 \\
-7x_2 - 6x_3 &= -10
\end{aligned}
$$

with the augmented matrix

$$
\begin{bmatrix} 1 & 2 & 1 & 3 \\ 0 & 1 & 1 & 2 \\ 0 & -7 & -6 & -10 \end{bmatrix}
$$

Now we replace the current third equation with the sum of the third and 7 multiple of the second,

$$
\begin{aligned}
x_1 + 2x_2 + x_3 &= 3 \\
x_2 + x_3 &= 2 \\
x_3 &= 4
\end{aligned}
$$

with the corresponding augmented matrix

$$\begin{bmatrix} 1 & 2 & 1 & 3 \\ 0 & 1 & 1 & 2 \\ 0 & 0 & 1 & 4 \end{bmatrix}$$

The system is triangular, which we can solve using the backward substitution. There is exactly one solution $(3, -2, 4)$ to the given linear system.

From the above example, we see that the elementary operations on equations of a linear system correspond to operations on the rows of the augmented matrix of the system. The **three elementary row operations** are (1) (**Replacement**) replace a row by its sum with a multiple of another row; (2) **(Interchange)** interchange any two rows; (3) **(Scaling)** multiply a row by a nonzero real number.

The three elementary row operations are applied to rows of matrices. During the process of reduction, one does not have to write names of all variables of a system.

1.1 Exercises

Exercise 9 Determine which equation is a linear equation in x_1, x_2, and x_3.

$$\text{(a) } \sqrt{2}x_1 - x_2 + 5x_3 = 1$$

$$\text{(b) } 4x_1 + x_2^{1/3} - x_3 = 2$$

$$\text{(c) } x_1 + 7x_2 + x_1x_3 = -6$$

$$\text{(d) } 2x_1 - \sqrt{3}x_2 + \pi x_3 = 0$$

Exercise 10 Provide the augmented matrix for the system of linear equations

$$\begin{aligned} x_1 + 6x_2 &= 0 \\ x_1 - x_2 &= 1 \\ 3x_1 + 4x_2 &= -5 \end{aligned}$$

Exercise 11 Provide the augmented matrix for the system of linear equations

$$\begin{aligned} x_1 + 6x_2 - 2x_3 &= 1 \\ x_1 - x_2 + x_3 &= 0 \\ 4x_3 &= -5 \end{aligned}$$

Exercise 12 Provide the augmented matrix for the system of linear equations

$$\begin{aligned} x_1 &= 3 \\ x_2 &= -8 \\ x_3 &= 0 \end{aligned}$$

Exercise 13 Find a system of linear equations with the given augmented matrix,

$$\begin{bmatrix} 3 & 0 & 1 \\ 6 & -5 & 0 \\ 0 & 2 & -1 \end{bmatrix}.$$

Exercise 14 Find a system of linear equations with the given augmented matrix,

$$\begin{bmatrix} 3 & 0 & 1 & 1 & 7 \\ 6 & -5 & 0 & 0 & 0 \\ 0 & 2 & -1 & 0 & -7 \end{bmatrix}.$$

Exercise 15 Solve the given system using elementary operations on the equations or elementary row operations on the augmented matrix.

$$\begin{aligned} x_1 - x_2 &= 5 \\ -3x_1 + x_2 &= 1 \end{aligned}$$

Exercise 16 Solve the given system using elementary operations on the equations or elementary row operations on the augmented matrix.

$$\begin{aligned} x_1 + x_2 - x_3 &= 5 \\ 2x_1 - x_2 + x_3 &= 1 \\ x_3 &= 2 \end{aligned}$$

Exercise 17 Solve the linear system whose augmented matrix is given as

$$\begin{bmatrix} 1 & 0 & 4 & 0 \\ 0 & 3 & 9 & 0 \\ 0 & 0 & 5 & -5 \end{bmatrix}.$$

1.2 Row Operations and Echelon Forms

Echelon Form

A system in a certain form, such as in triangular form, can be relatively easy to solve. In this section, echelon form and reduced echelon form of a matrix are introduced. These forms are friendly and easy to solve due to the zero entries appearing in the matrix. In a row having nonzero entries, the leftmost nonzero entry is called the leading entry of the row. A matrix is in **echelon form** (or **row echelon form**) if

(1) the rows with only zero entries are below the rows having nonzero entries;

(2) the leading entry of a row lies to the right of the leading entry of the row above;

(3) all entries below a leading entry are zeros in that column.

Example 1 (*Echelon Form*) The following matrices are in echelon form.

$$\begin{bmatrix} 2 & 3 & 4 \\ 0 & 1 & 6 \\ 0 & 0 & 5 \end{bmatrix}, \quad \begin{bmatrix} 3 & 1 & 2 \\ 0 & 0 & 1 \\ 0 & 0 & 0 \end{bmatrix}, \quad \begin{bmatrix} 2 & 6 & 3 & 1 & 0 \\ 0 & 0 & 2 & 6 & 1 \\ 0 & 0 & 0 & 0 & 0 \end{bmatrix}$$

A matrix is in **reduced echelon form** (or **reduced row echelon form**) if it satisfies the following additional conditions:

(4) The leading entry is 1 for any nonzero row;

(5) Any leading entry 1 is the only nonzero entry in its column.

Example 2 (*Reduced Echelon Form*) The following matrices are in reduced echelon form

$$\begin{bmatrix} 1 & 0 & 0 \\ 0 & 1 & 0 \\ 0 & 0 & 0 \end{bmatrix}, \quad \begin{bmatrix} 1 & 0 & 0 & 2 \\ 0 & 1 & 0 & 0 \\ 0 & 0 & 1 & 3 \end{bmatrix}, \quad \begin{bmatrix} 1 & 8 & 0 & 1 & 2 \\ 0 & 0 & 1 & -3 & 1 \\ 0 & 0 & 0 & 0 & 0 \end{bmatrix}$$

Remark 3 It can be proved that each matrix is row equivalent to one and only one reduced echelon form. But this is not true for echelon form. A matrix can be row equivalent to more than one echelon form. For example, the following three matrices in echelon form are row equivalent

$$\begin{bmatrix} 2 & 3 & 1 \\ 0 & 2 & 4 \\ 0 & 0 & 0 \end{bmatrix}, \quad \begin{bmatrix} 2 & 3 & 1 \\ 0 & 1 & 2 \\ 0 & 0 & 0 \end{bmatrix}, \quad \begin{bmatrix} 2 & 0 & -5 \\ 0 & 1 & 2 \\ 0 & 0 & 0 \end{bmatrix}.$$

The echelon form of a matrix is not unique. But the reduced echelon form of a matrix is unique.

The echelon form or reduced echelon form of the augmented matrix of a linear system can help us see the answers to the following questions:

(i) Is the linear system consistent? That is, does the system have at least one solution?
(ii) If the linear system is consistent, is the solution unique?

Example 4 (*Using Echelon Forms*) Determine if the linear system is consistent using the echelon form of its augmented matrix

$$\begin{bmatrix} 1 & 3 & 5 & -2 \\ 0 & 6 & 4 & -5 \\ 0 & 0 & 0 & 2 \end{bmatrix}.$$

The third row of the echelon form corresponds to the equation $0 = 2$, which is false no matter what values of the variables are. The system has no solution. It is inconsistent.

Example 5 (*Using Echelon Forms*) Determine if the linear system is consistent using the reduced echelon form of its augmented matrix

$$\begin{bmatrix} 1 & 0 & 0 & 0 & 1 \\ 0 & 0 & 1 & 0 & 6 \\ 0 & 0 & 0 & 1 & 4 \end{bmatrix}.$$

The reduced echelon form corresponds to the system of three equations involving four variables,

$$\begin{aligned} x_1 &= 1 \\ x_3 &= 6 \\ x_4 &= 4 \end{aligned}$$

Since x_2 can take any real number t, the linear system has infinite number of solutions of the form $(1, t, 6, 4)$, with t being an arbitrary real number. Hence the given system is consistent.

Example 6 (*Using Echelon Forms*) Determine if the linear system is consistent using the reduced echelon form of its augmented matrix

$$\begin{bmatrix} 1 & 0 & 0 & 2 \\ 0 & 1 & 0 & 5 \\ 0 & 0 & 1 & 7 \end{bmatrix}.$$

The reduced echelon form corresponds to the system of three equations in three variables,

$$\begin{aligned} x_1 &= 2 \\ x_2 &= 5 \\ x_3 &= 7 \end{aligned}$$

This system is consistent. Its solution (2, 5, 7) is unique.

Row Reduction

A **pivot position** of a matrix is the position where a leading 1 is in its reduced echelon form. A **pivot** is a nonzero entry in a pivot position. A **pivot column** is a column containing a pivot position.

Example 7 (*Pivot Positions and Pivot Columns*) Consider the matrix

$$A = \begin{bmatrix} 0 & 0 & 1 & 8 \\ 0 & 0 & 0 & 0 \\ 1 & 0 & 0 & 9 \end{bmatrix}.$$

Row reduce the matrix A by interchanging the first row and the third row,

$$\begin{bmatrix} 1 & 0 & 0 & 9 \\ 0 & 0 & 0 & 0 \\ 0 & 0 & 1 & 8 \end{bmatrix},$$

Next we interchange the second row of zeros and the third row to get the reduced echelon form,

$$\begin{bmatrix} 1 & 0 & 0 & 9 \\ 0 & 0 & 1 & 8 \\ 0 & 0 & 0 & 0 \end{bmatrix}.$$

We see that the first pivot position of A is where the leftmost leading 1 is, at the top of the first column. The other pivot position of A is where the other leading 1 lies, which is at the second row and third column. The pivot columns are the first column and the third column of A since each of the two columns contains a pivot position.

Example 8 (*Row Reduction and Solve a System*) Solve the system of equations:

$$\begin{aligned} x_1 + 2x_2 + 3x_3 &= 9 \\ 2x_1 + 5x_2 + 7x_3 &= 20 \\ -2x_3 &= -6 \end{aligned}$$

Let's use elementary row operations to reduce its augmented matrix

$$\begin{bmatrix} 1 & 2 & 3 & 9 \\ 2 & 5 & 7 & 20 \\ 0 & 0 & -2 & -6 \end{bmatrix}$$

into the reduced echelon form. The nonzero entry 1 in the first row is a pivot. Adding -2 times row 1 to row 2, we replace row 2. The resulting matrix is in echelon form,

$$\begin{bmatrix} 1 & 2 & 3 & 9 \\ 0 & 1 & 1 & 2 \\ 0 & 0 & -2 & -6 \end{bmatrix}, \text{ or } \begin{array}{rrrl} x_1 & +2x_2 & +3x_3 & = 9 \\ & x_2 & +x_3 & = 2 \\ & & -2x_3 & = -6 \end{array}.$$

Let's continue to reduce the matrix to reduced echelon form. The second pivot is the leading 1 of the second row. By adding -2 times row 2 to row 1, we replace row 1 to obtain

$$\begin{bmatrix} 1 & 0 & 1 & 5 \\ 0 & 1 & 1 & 2 \\ 0 & 0 & -2 & -6 \end{bmatrix}, \text{ or } \begin{array}{rrrl} x_1 & & +x_3 & = 5 \\ & x_2 & +x_3 & = 2 \\ & & -2x_3 & = -6 \end{array}.$$

The next pivot is -2 in row 3. Let's first scale row 3 by multiplying its entries by $-1/2$. This is to have 1 as the coefficient for x_3 in equation 3,

$$\begin{bmatrix} 1 & 0 & 1 & 5 \\ 0 & 1 & 1 & 2 \\ 0 & 0 & 1 & 3 \end{bmatrix}, \text{ or } \begin{array}{rrrl} x_1 & & +x_3 & = 5 \\ & x_2 & +x_3 & = 2 \\ & & x_3 & = 3 \end{array}.$$

Then we replace row 1 by subtracting row 3 from row 1, and replace row 2 by subtracting row 3 from row 2,

$$\begin{bmatrix} 1 & 0 & 0 & 2 \\ 0 & 1 & 0 & -1 \\ 0 & 0 & 1 & 3 \end{bmatrix}, \text{ or } \begin{array}{rrrl} x_1 & & & = 2 \\ & x_2 & & = -1 \\ & & x_3 & = 3 \end{array}.$$

The reduced echelon form reveals the unique solution $(2, -1, 3)$ of the linear system.

Example 9 (*Row Reduction and Solve a System*) Solve the system of equations:

$$\begin{aligned} x_1 - 3x_2 - 5x_3 &= 0 \\ x_2 + x_3 &= 3 \end{aligned}$$

Let's use elementary row operations to reduce its augmented matrix

$$\begin{bmatrix} 1 & -3 & -5 & 0 \\ 0 & 1 & 1 & 3 \end{bmatrix}$$

into the reduced echelon form. The nonzero entry 1 in the first row is a pivot. The column 1, the pivot column containing this pivot, has only zero entries other than the pivot. Now we identify the next pivot, the leading 1 in the second row. Adding 3 times row 2 to row 1, we replace row 1. The resulting matrix is now in reduced echelon form,

$$\begin{bmatrix} 1 & 0 & -2 & 9 \\ 0 & 1 & 1 & 3 \end{bmatrix}, \text{ or } \begin{array}{rl} x_1 \quad -2x_3 &= 9 \\ x_2 \ +x_3 &= 3 \end{array}.$$

We have

$$x_1 = 9 + 2x_3$$
$$x_2 = 3 - x_3$$

with x_3 being an arbitrary real number t. For examples, $(9,\ 3,\ 0)$ is a solution when $x_3 = 0$, and $(9 + 2\pi,\ 3 - \pi,\ \pi)$ is another solution when $x_3 = \pi$. But we should list all possible solutions, which are infinitely many. Any solution of the system can be expressed in the form $(9 + 2x_3,\ 3 - x_3,\ x_3)$, with x_3 being an arbitrary real number. Here, x_3 is called a free variable. The other two variables x_1 and x_2 correspond to pivot columns are called basic variables. We could also use t (instead of x_3) for any real number and express the general solutions as $(9 + 2t,\ 3 - t,\ t)$.

Example 10 (*Row Reduction and Solve a System*) Solve the system of equations:

$$2x + 4y - 2z = 2$$
$$x + 2y - z = 5$$

Let's use elementary row operations to reduce its augmented matrix

$$\begin{bmatrix} 2 & 4 & -2 & 2 \\ 1 & 2 & -1 & 5 \end{bmatrix}$$

into the reduced echelon form. The entry 2 in column 1 is in pivot position. To have a leading 1 as a pivot, we can either scale row 1 or interchange rows 1 and 2. Let's scale row 1 by multiplying all entries of row 1 by 1/2. We have

$$\begin{bmatrix} 1 & 2 & -1 & 1 \\ 1 & 2 & -1 & 5 \end{bmatrix}.$$

Now we replace row 2 by subtracting from row 2 the first row,

$$\begin{bmatrix} 1 & 2 & -1 & 1 \\ 0 & 0 & 0 & 4 \end{bmatrix}.$$

Then scaling row 2 by multiplying all its entries by 1/4, we obtain the reduced echelon form,

$$\begin{bmatrix} 1 & 2 & -1 & 0 \\ 0 & 0 & 0 & 1 \end{bmatrix}$$

which corresponds to the system of equations

$$\begin{aligned} x + 2y - z &= 0 \\ 0 &= 1 \end{aligned}$$

Hence, there is no solution to the system.

1.2 Exercises

Exercise 11 Determine which of the following matrices are in echelon form and which ones are in reduced echelon form.

$$\text{a.} \begin{bmatrix} 1 & 0 & -1 & 5 \\ 0 & 2 & 1 & 0 \\ 0 & 0 & 1 & 3 \end{bmatrix} \quad \text{b.} \begin{bmatrix} 1 & 2 & -1 & 5 \\ 0 & 0 & 0 & 0 \end{bmatrix}$$

$$\text{c.} \begin{bmatrix} 1 & 0 & 0 & 0 \\ 0 & 0 & 1 & 0 \\ 0 & 0 & 0 & 6 \end{bmatrix} \quad \text{d.} \begin{bmatrix} 1 & 0 & 3 & 0 & 8 \\ 0 & 1 & 2 & 0 & -3 \\ 0 & 0 & 0 & 1 & 2 \end{bmatrix}$$

Exercise 12 Use row reduction to solve each system of equations.

$$\text{a.} \begin{aligned} x + 2y - z &= 1 \\ x + 2y \phantom{{}-z} &= 5 \end{aligned}$$

$$\text{b.} \begin{aligned} 2y + 2z &= 0 \\ 2x - y + 3z &= -2 \\ -4x + 4y - 4z &= 2 \end{aligned}$$

Exercise 13 Find the solution set for each linear system whose augmented matrix has been transformed into reduced echelon form.

$$\text{a.} \begin{bmatrix} 1 & 0 & 0 & -1 & 6 \\ 0 & 0 & 1 & 2 & 3 \end{bmatrix}$$

$$\text{b.} \begin{bmatrix} 1 & 0 & 0 & 2 \\ 0 & 1 & 0 & 8 \\ 0 & 0 & 0 & 0 \end{bmatrix}$$

Exercise 14 Use row reduction to solve each linear system whose augmented matrix given as below.

$$\text{a.} \begin{bmatrix} 2 & 0 & -2 & 0 \\ 3 & 1 & 0 & 0 \\ 2 & 3 & 4 & 5 \end{bmatrix}$$

b. $\begin{bmatrix} 1 & 2 & -1 & 4 \\ 2 & 4 & 3 & 5 \\ 1 & 2 & -6 & 7 \end{bmatrix}$

c. $\begin{bmatrix} 1 & 2 & 1 & 2 & 1 & 2 \\ 0 & 0 & 1 & -1 & -1 & 4 \\ 2 & 4 & 3 & 3 & 3 & 4 \\ 3 & 6 & 6 & 3 & 6 & 6 \end{bmatrix}$

1.3 Vector Equations

Vectors in $\mathbf{R}^2$

A matrix with only one column is called a column vector, or simply a vector. We begin this section with vectors of two entries. The vectors with two entries in $\mathbf{R}^2$ provide us a familiar setting and the convenience for geometric descriptions. A point in the Cartesian plane is denoted by an ordered pair of real numbers (x_1, x_2) . The arrow pointing from the origin (0, 0) to the point (x_1, x_2) defines a vector $\begin{bmatrix} x_1 \\ x_2 \end{bmatrix}$ in $\mathbf{R}^2$ The vector has its length and direction. Two vectors are equal if and only if they have the same length and the same direction. Algebraically, two vectors in $\mathbf{R}^2$ are equal if and only if their corresponding entries are equal. The sum $\mathbf{x} + \mathbf{y}$ of two vectors $\mathbf{x}$ and $\mathbf{y}$ in $\mathbf{R}^2$ is obtained by adding corresponding entries of $\mathbf{x}$ and $\mathbf{y}$. For a vector $\mathbf{x}$ in $\mathbf{R}^2$ and a real number c, the vector $c\mathbf{x}$, called the scalar multiple of $\mathbf{x}$ by c, is obtained by multiplying each entry of $\mathbf{x}$ by c.

Example 1 (*Vector Operations*) Given $\mathbf{x} = \begin{bmatrix} 1 \\ 6 \end{bmatrix}$ and $\mathbf{y} = \begin{bmatrix} 4 \\ -3 \end{bmatrix}$, we can find $\mathbf{x} + \mathbf{y}$, $3\mathbf{x}$, $(-2)\mathbf{y}$, and $3\mathbf{x}+(-2)\mathbf{y}$ as follows:

$$\mathbf{x} + \mathbf{y} = \begin{bmatrix} 1 \\ 6 \end{bmatrix} + \begin{bmatrix} 4 \\ -3 \end{bmatrix} = \begin{bmatrix} 5 \\ 3 \end{bmatrix}$$

$$3\mathbf{x} = 3\begin{bmatrix} 1 \\ 6 \end{bmatrix} = \begin{bmatrix} 3 \\ 18 \end{bmatrix}$$

$$(-2)\,\mathbf{y} = (-2)\begin{bmatrix} 4 \\ -3 \end{bmatrix} = \begin{bmatrix} -8 \\ 6 \end{bmatrix}$$

$$3\mathbf{x}+(-2)\mathbf{y} = \begin{bmatrix} 3 \\ 18 \end{bmatrix} + \begin{bmatrix} -8 \\ 6 \end{bmatrix} = \begin{bmatrix} -5 \\ 24 \end{bmatrix}$$

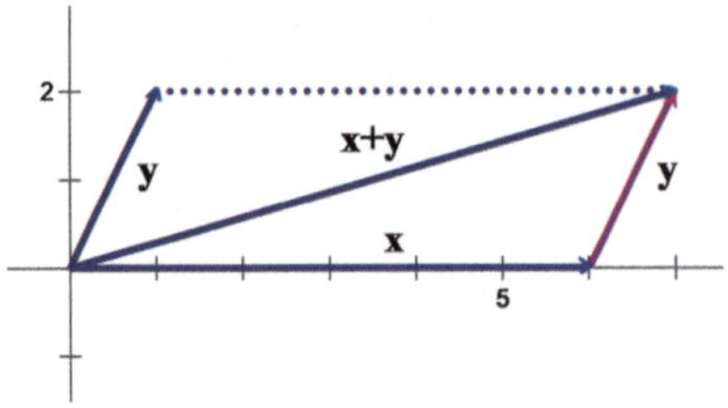

Fig. 1 Parallelogram rule for vector addition

Remark 2 In above example, the vector $3\mathbf{x}$ is three times as long as the vector $\mathbf{x}$ with its direction being the same as $\mathbf{x}'s$; The vector $(-2)\,\mathbf{y}$ is twice as long as the vector $\mathbf{y}$ with direction being opposite of $\mathbf{y}'s$; The vector $\mathbf{x}+\mathbf{y}$ is the "long" diagonal of the parallelogram formed by $\mathbf{x}$ and $\mathbf{y}$ (see Fig. 1 for an illustration of $\mathbf{x}+\mathbf{y}$).

Vectors in $\mathbf{R}^n$

Vectors in $\mathbf{R}^n$ are column vectors with n entries,

$$\mathbf{x} = \begin{bmatrix} x_1 \\ x_2 \\ \vdots \\ x_n \end{bmatrix}.$$

For example, vectors in $\mathbf{R}^3$ are column vectors with 3 entries,

$$\mathbf{x} = \begin{bmatrix} x_1 \\ x_2 \\ x_3 \end{bmatrix}.$$

Each vector in $\mathbf{R}^n$, like a vector in $\mathbf{R}^2$, has its length and direction. The zero vector $\mathbf{0}$ in $\mathbf{R}^n$ is the vector whose entries are all zero. This is the only vector that has no direction. Two vectors are **equal** if and only if they have the same length and the same direction. Algebraically, two vectors in $\mathbf{R}^n$ are equal if and only if their corresponding entries are equal. The **sum** $\mathbf{x}+\mathbf{y}$ of two vectors $\mathbf{x}$ and $\mathbf{y}$ in $\mathbf{R}^n$ is obtained by adding corresponding entries of $\mathbf{x}$ and $\mathbf{y}$. For a vector $\mathbf{x}$ in $\mathbf{R}^n$ and a real number c, the vector $c\mathbf{x}$, called the **scalar multiple** of $\mathbf{x}$ by c, is obtained by multiplying each entry of $\mathbf{x}$ by c.

Example 3 (*Vector Operations*) Given two vectors in $\mathbf{R}^3$,

$$\mathbf{u} = \begin{bmatrix} 1 \\ 2 \\ 3 \end{bmatrix} \text{ and } \mathbf{v} = \begin{bmatrix} -2 \\ 6 \\ 4 \end{bmatrix}.$$

We can determine $\mathbf{u} + \mathbf{v}$, $-2\mathbf{u}$, and $\mathbf{u}+\frac{1}{2}\mathbf{v}$,

$$\mathbf{u} + \mathbf{v} = \begin{bmatrix} 1 \\ 2 \\ 3 \end{bmatrix} + \begin{bmatrix} -2 \\ 6 \\ 4 \end{bmatrix} = \begin{bmatrix} -1 \\ 8 \\ 7 \end{bmatrix}$$

$$-2\mathbf{u} = (-2) \begin{bmatrix} 1 \\ 2 \\ 3 \end{bmatrix} = \begin{bmatrix} -2 \\ -4 \\ -6 \end{bmatrix}$$

$$\mathbf{u}+\frac{1}{2}\mathbf{v} = \begin{bmatrix} 1 \\ 2 \\ 3 \end{bmatrix} + \frac{1}{2} \begin{bmatrix} -2 \\ 6 \\ 4 \end{bmatrix} = \begin{bmatrix} 1 \\ 2 \\ 3 \end{bmatrix} + \begin{bmatrix} -1 \\ 3 \\ 2 \end{bmatrix} = \begin{bmatrix} 0 \\ 5 \\ 5 \end{bmatrix}$$

Remark 4 In the above example, we generated from the given vectors **u** and **v** the new vectors, such as $\mathbf{u} + \mathbf{v}$, $-2\mathbf{u}$, and $\mathbf{u}+\frac{1}{2}\mathbf{v}$. But many more vectors in the form of $a\mathbf{u}+b\mathbf{v}$, with a and b being real numbers, can be generated by **u** and **v**.

The following algebraic properties can be proved true for all $\mathbf{x}$, $\mathbf{y}$, $\mathbf{z}$ in $\mathbf{R}^n$ and all scalars c and d :

(a) $\mathbf{x} + \mathbf{y} = \mathbf{y} + \mathbf{x}$
(b) $(\mathbf{x} + \mathbf{y}) + \mathbf{z} = \mathbf{x} + (\mathbf{y} + \mathbf{z})$
(c) $\mathbf{0} + \mathbf{x} = \mathbf{x} + \mathbf{0} = \mathbf{x}$
(d) $\mathbf{x} + (-\mathbf{x}) = -\mathbf{x} + \mathbf{x} = \mathbf{0}$, where $-\mathbf{x}$ is the same as $(-1)\,\mathbf{x}$.
(e) $c\,(\mathbf{x} + \mathbf{y}) = c\mathbf{x} + c\mathbf{y}$
(f) $(c + d)\,\mathbf{x} = c\mathbf{x} + d\mathbf{x}$
(g) $c\,(d\mathbf{x}) = (cd)\,\mathbf{x}$
(h) $1\mathbf{x} = \mathbf{x}$

Linear Combinations

The vector **u** is called a **linear combination** of the vectors $\mathbf{v}_1, \mathbf{v}_2, \ldots, \mathbf{v}_p$ in $\mathbf{R}^n$ if **u** can be represented by the p vectors as,

$$\mathbf{u} = c_1\mathbf{v}_1 + c_2\mathbf{v}_2 + \cdots + c_p\mathbf{v}_p$$

where $c_1, c_2, \ldots, c_p$ are scalars.

Example 5 (*Linear Combinations*) Some linear combinations of vectors $\mathbf{e}_1 = \begin{bmatrix} 1 \\ 0 \end{bmatrix}$ and $\mathbf{e}_2 = \begin{bmatrix} 0 \\ 1 \end{bmatrix}$ in $\mathbf{R}^2$ are:

$$2\begin{bmatrix} 1 \\ 0 \end{bmatrix} + 5\begin{bmatrix} 0 \\ 1 \end{bmatrix} = \begin{bmatrix} 2 \\ 5 \end{bmatrix}$$

$$10\begin{bmatrix} 1 \\ 0 \end{bmatrix} + (-10)\begin{bmatrix} 0 \\ 1 \end{bmatrix} = \begin{bmatrix} 10 \\ -10 \end{bmatrix}$$

$$0\begin{bmatrix} 1 \\ 0 \end{bmatrix} + (-8)\begin{bmatrix} 0 \\ 1 \end{bmatrix} = \begin{bmatrix} 0 \\ -8 \end{bmatrix}$$

$$0\begin{bmatrix} 1 \\ 0 \end{bmatrix} + 0\begin{bmatrix} 0 \\ 1 \end{bmatrix} = \begin{bmatrix} 0 \\ 0 \end{bmatrix}$$

That is, the vector $\begin{bmatrix} 2 \\ 5 \end{bmatrix}$ is a linear combination of $\mathbf{e}_1$ and $\mathbf{e}_2$ with weights 2 and 5; the vector $\begin{bmatrix} 0 \\ -8 \end{bmatrix}$ is a linear combination of $\mathbf{e}_1$ and $\mathbf{e}_2$ with weights 0 and -8; the vector $\begin{bmatrix} 0 \\ 0 \end{bmatrix}$ is a linear combination of $\mathbf{e}_1$ and $\mathbf{e}_2$ with weights 0 and 0;

Example 6 (*Linear Combinations*) Determine if the vector $\mathbf{u} = \begin{bmatrix} 8 \\ 19 \\ 11 \end{bmatrix}$ can be represented as a linear combination $\mathbf{v}_1 = \begin{bmatrix} 1 \\ 2 \\ 4 \end{bmatrix}$ and $\mathbf{v}_2 = \begin{bmatrix} 2 \\ 5 \\ 1 \end{bmatrix}$. Here we want to know if there are possible values of weights x_1 and x_2 such that

$$x_1\mathbf{v}_1 + x_2\mathbf{v}_2 = \mathbf{u},$$

i.e.,

$$x_1\begin{bmatrix} 1 \\ 2 \\ 4 \end{bmatrix} + x_2\begin{bmatrix} 2 \\ 5 \\ 1 \end{bmatrix} = \begin{bmatrix} 8 \\ 19 \\ 11 \end{bmatrix}.$$

Let's rewrite the vector equation as

$$\begin{bmatrix} x_1 \\ 2x_1 \\ 4x_1 \end{bmatrix} + \begin{bmatrix} 2x_2 \\ 5x_2 \\ x_2 \end{bmatrix} = \begin{bmatrix} 8 \\ 19 \\ 11 \end{bmatrix},$$

and

$$\begin{bmatrix} x_1 + 2x_2 \\ 2x_1 + 5x_2 \\ 4x_1 + x_2 \end{bmatrix} = \begin{bmatrix} 8 \\ 19 \\ 11 \end{bmatrix}.$$

The above two vectors are equal if and only if x_1 and x_2 satisfy the following system:

$$\begin{aligned} x_1 + 2x_2 &= 8 \\ 2x_1 + 5x_2 &= 19 \\ 4x_1 + x_2 &= 11 \end{aligned}$$

By reducing the augmented matrix

$$\begin{bmatrix} 1 & 2 & 8 \\ 2 & 5 & 19 \\ 4 & 1 & 11 \end{bmatrix} \sim \begin{bmatrix} 1 & 2 & 8 \\ 0 & 1 & 3 \\ 0 & -7 & -21 \end{bmatrix} \sim \begin{bmatrix} 1 & 0 & 2 \\ 0 & 1 & 3 \\ 0 & 0 & 0 \end{bmatrix}$$

which indicates that the system is consistent and it has the solution $x_1 = 2$ and $x_2 = 3$. So the vector **u** can be generated by $\mathbf{v}_1$ and $\mathbf{v}_2$ with weights 2 and 3, that is, $\mathbf{u} = 2\mathbf{v}_1 + 3\mathbf{v}_2$.

The vector **u** in $\mathbf{R}^n$ can be generated by the vectors $\mathbf{v}_1, \mathbf{v}_2, \ldots, \mathbf{v}_p$ in $\mathbf{R}^n$ when the following **vector equation**

$$x_1\mathbf{v}_1 + x_2\mathbf{v}_2 + \cdots + x_p\mathbf{v}_p = \mathbf{u}$$

has at least one solution. We know that the vector equation has the same solution set as the system whose augmented matrix is

$$\begin{bmatrix} \mathbf{v}_1 & \mathbf{v}_2 & \cdots & \mathbf{v}_p & \mathbf{u} \end{bmatrix}$$

The vector **u** can be generated by $\mathbf{v}_1, \mathbf{v}_2, \ldots, \mathbf{v}_p$ if and only if the system corresponding to the above matrix is consistent.

The set of all linear combinations of the vectors $\mathbf{v}_1, \mathbf{v}_2, \ldots, \mathbf{v}_p$ in $\mathbf{R}^n$ is called the subset of $\mathbf{R}^n$ spanned by $\mathbf{v}_1, \mathbf{v}_2, \ldots, \mathbf{v}_p$, and the set is denoted by $Span\ \{\mathbf{v}_1, \mathbf{v}_2, \ldots, \mathbf{v}_p\}$.

Example 7 (*A Subset Spanned by Given Vectors*) Let's consider the vectors $\mathbf{e}_1 = \begin{bmatrix} 1 \\ 0 \end{bmatrix}$ and $\mathbf{e}_2 = \begin{bmatrix} 0 \\ 1 \end{bmatrix}$ in $\mathbf{R}^2$ again. As a matter of fact, the two vectors $\mathbf{e}_1$ and $\mathbf{e}_2$ can generate any vector $\begin{bmatrix} a \\ b \end{bmatrix}$ in $\mathbf{R}^2$:

$$a\begin{bmatrix} 1 \\ 0 \end{bmatrix} + b\begin{bmatrix} 0 \\ 1 \end{bmatrix} = \begin{bmatrix} a \\ b \end{bmatrix}$$

That is, the set of all vectors in $\mathbf{R}^2$ is the same as $Span\ \{\mathbf{e}_1, \mathbf{e}_2\}$.

1.3 Exercises

Exercise 8 Sketch the two vectors $\mathbf{u} = \begin{bmatrix} 1 \\ 0 \end{bmatrix}$ and $\mathbf{v} = \begin{bmatrix} 1 \\ 1 \end{bmatrix}$ in $\mathbf{R}^2$. Sketch the vectors $-\mathbf{u}$, $-2\mathbf{v}$, $2\mathbf{u}$, $2\mathbf{v}$, $3\mathbf{u}$, $3\mathbf{v}$, $2\mathbf{u}+2\mathbf{v}$, $3\mathbf{u}-2\mathbf{v}$, $2\mathbf{u}+3\mathbf{v}$, $3\mathbf{u}+3\mathbf{v}$ on the same graph.

Exercise 9 Determine if the vector $\begin{bmatrix} 2 \\ 5 \end{bmatrix}$ can be generated by $\mathbf{u} = \begin{bmatrix} 1 \\ 0 \end{bmatrix}$ and $\mathbf{v} = \begin{bmatrix} 1 \\ 1 \end{bmatrix}$. Justify your answer.

Exercise 10 Determine if any arbitrary vector $\begin{bmatrix} a \\ b \end{bmatrix}$ in $\mathbf{R}^2$ can be generated by $\mathbf{u} = \begin{bmatrix} 1 \\ 0 \end{bmatrix}$ and $\mathbf{v} = \begin{bmatrix} 1 \\ 1 \end{bmatrix}$. Justify your answer.

Exercise 11 Determine if the vector $\mathbf{w} = \begin{bmatrix} 0 \\ -1 \\ 7 \end{bmatrix}$ can be written as a linear combination of $\mathbf{v}_1 = \begin{bmatrix} 1 \\ 2 \\ 4 \end{bmatrix}$ and $\mathbf{v}_2 = \begin{bmatrix} 2 \\ 5 \\ 1 \end{bmatrix}$.

Exercise 12 Determine if the vector $\mathbf{b} = \begin{bmatrix} 2 \\ -1 \\ 6 \end{bmatrix}$ is in the subset spanned by the vectors $\mathbf{a}_1 = \begin{bmatrix} 1 \\ -2 \\ 0 \end{bmatrix}$, $\mathbf{a}_2 = \begin{bmatrix} 0 \\ 1 \\ 2 \end{bmatrix}$, and $\mathbf{a}_3 = \begin{bmatrix} 5 \\ -6 \\ 8 \end{bmatrix}$.

Exercise 13 Given $\mathbf{e}_1 = \begin{bmatrix} 1 \\ 0 \\ 0 \end{bmatrix}$, $\mathbf{e}_2 = \begin{bmatrix} 0 \\ 1 \\ 0 \end{bmatrix}$, and $\mathbf{e}_3 = \begin{bmatrix} 0 \\ 0 \\ 1 \end{bmatrix}$ in $\mathbf{R}^3$. Determine if $\mathbf{b} = \begin{bmatrix} 2 \\ -1 \\ 6 \end{bmatrix}$ is in the subset spanned by $\mathbf{e}_1$, $\mathbf{e}_2$, and $\mathbf{e}_3$.

Exercise 14 Determine if any arbitrary vector $\begin{bmatrix} a \\ b \\ c \end{bmatrix}$ in $\mathbf{R}^3$ can be generated by $\mathbf{e}_1 = \begin{bmatrix} 1 \\ 0 \\ 0 \end{bmatrix}$, $\mathbf{e}_2 = \begin{bmatrix} 0 \\ 1 \\ 0 \end{bmatrix}$, and $\mathbf{e}_3 = \begin{bmatrix} 0 \\ 0 \\ 1 \end{bmatrix}$. Justify your answer.

1.4 Matrix Equations

Matrix Multiplication

The simplest linear system is a system of one equation in one variable x, in the form

$$ax = b$$

where a, b, and x are scalars. We can generalize the form to

$$A\mathbf{x} = \mathbf{b}$$

to represent a system of m equations and n variables, where A is an $m \times n$ matrix, $\mathbf{x}$ is in $\mathbf{R}^n$, and $\mathbf{b}$ is in $\mathbf{R}^m$. Consider the system:

$$\begin{aligned} a_{11}x_1 + a_{12}x_2 + \cdots + a_{1n}x_n &= b_1 \\ a_{21}x_1 + a_{22}x_2 + \cdots + a_{2n}x_n &= b_2 \\ &\vdots \\ a_{m1}x_1 + a_{m2}x_2 + \cdots + a_{mn}x_n &= b_m \end{aligned}$$

We can write it as a matrix equation in the form $A\mathbf{x} = \mathbf{b}$ if we set

$$A = \begin{bmatrix} a_{11} & a_{12} & \cdots & a_{1n} \\ a_{21} & a_{22} & \cdots & a_{2n} \\ \vdots & & & \\ a_{m1} & a_{m2} & \cdots & a_{mn} \end{bmatrix}, \quad \mathbf{x} = \begin{bmatrix} x_1 \\ x_2 \\ \vdots \\ x_n \end{bmatrix}, \quad \mathbf{b} = \begin{bmatrix} b_1 \\ b_2 \\ \vdots \\ b_m \end{bmatrix}$$

and define **the product** $A\mathbf{x}$ as

$$A\mathbf{x} = \begin{bmatrix} a_{11}x_1 + a_{12}x_2 + \cdots + a_{1n}x_n \\ a_{21}x_1 + a_{22}x_2 + \cdots + a_{2n}x_n \\ \vdots \\ a_{m1}x_1 + a_{m2}x_2 + \cdots + a_{mn}x_n \end{bmatrix}.$$

Example 1 (*A Matrix Equation*) Let $A = \begin{bmatrix} 1 & 2 & 3 \\ 4 & 5 & 6 \end{bmatrix}$, $\mathbf{x} = \begin{bmatrix} x_1 \\ x_2 \\ x_3 \end{bmatrix}$, and $\mathbf{b} = \begin{bmatrix} 8 \\ -9 \end{bmatrix}$. The matrix equation

$$A\mathbf{x} = \mathbf{b}, \quad \text{i.e.,} \quad \begin{bmatrix} 1 & 2 & 3 \\ 4 & 5 & 6 \end{bmatrix} \begin{bmatrix} x_1 \\ x_2 \\ x_3 \end{bmatrix} = \begin{bmatrix} 8 \\ -9 \end{bmatrix}$$

represents the linear system

$$\begin{bmatrix} x_1 + 2x_2 + 3x_3 \\ 4x_1 + 5x_2 + 6x_3 \end{bmatrix} = \begin{bmatrix} 8 \\ -9 \end{bmatrix}$$

or

$$\begin{aligned} x_1 + 2x_2 + 3x_3 &= 8 \\ 4x_1 + 5x_2 + 6x_3 &= -9 \end{aligned}$$

We realize that $A\mathbf{x}$ is a vector, of the same dimension as $\mathbf{b}$, which is a column of 2 entries. Here, A is a 2×3 matrix, and the vector $\mathbf{x}$ is a column vector of 3 entries. For $A\mathbf{x}$ to be a valid operation, the number of entries of $\mathbf{x}$ must be equal to the number of columns of A. The vector $A\mathbf{x}$ is a linear combination of the columns of A :

$$\begin{bmatrix} x_1 + 2x_2 + 3x_3 \\ 4x_1 + 5x_2 + 6x_3 \end{bmatrix} = x_1 \begin{bmatrix} 1 \\ 4 \end{bmatrix} + x_2 \begin{bmatrix} 2 \\ 5 \end{bmatrix} + x_3 \begin{bmatrix} 3 \\ 6 \end{bmatrix}$$

Example 2 (*Product of a Matrix and a Vector*) Let $A = \begin{bmatrix} 3 & 1 \\ 0 & -4 \\ 1 & 0 \\ 4 & 5 \end{bmatrix}$, $\mathbf{x} = \begin{bmatrix} x_1 \\ x_2 \end{bmatrix}$, and $\mathbf{u} = \begin{bmatrix} 2 \\ -3 \end{bmatrix}$, then

$$A\mathbf{x} = \begin{bmatrix} 3 & 1 \\ 0 & -4 \\ 1 & 0 \\ 4 & 5 \end{bmatrix} \begin{bmatrix} x_1 \\ x_2 \end{bmatrix} = x_1 \begin{bmatrix} 3 \\ 0 \\ 1 \\ 4 \end{bmatrix} + x_2 \begin{bmatrix} 1 \\ -4 \\ 0 \\ 5 \end{bmatrix}$$

$$A\mathbf{u} = \begin{bmatrix} 3 & 1 \\ 0 & -4 \\ 1 & 0 \\ 4 & 5 \end{bmatrix} \begin{bmatrix} 2 \\ -3 \end{bmatrix} = 2 \begin{bmatrix} 3 \\ 0 \\ 1 \\ 4 \end{bmatrix} + (-3) \begin{bmatrix} 1 \\ -4 \\ 0 \\ 5 \end{bmatrix} = \begin{bmatrix} 3 \\ 12 \\ 2 \\ -7 \end{bmatrix}$$

We notice that the matrix A in this example is of 4×2, that is, the matrix has 2 columns with each column vector in R^4. The vector $\mathbf{x}$ (or $\mathbf{u}$) is a column vector of 2 entries, the same as the number of columns of A. The resulted $A\mathbf{x}$ (or $A\mathbf{u}$) is in R^4 since $A\mathbf{x}$ (or $A\mathbf{u}$) is a linear combination of column vectors of A.

Equivalence of Different Forms

The same linear system may appear in different forms.

Example 3 (*Linear System in Different Forms*) Let's determine if the vector $\mathbf{b} = \begin{bmatrix} 3 \\ 12 \\ 2 \\ -7 \end{bmatrix}$ is in the subset spanned by $\mathbf{a}_1 = \begin{bmatrix} 3 \\ 0 \\ 1 \\ 4 \end{bmatrix}$ and $\mathbf{a}_2 = \begin{bmatrix} 1 \\ -4 \\ 0 \\ 5 \end{bmatrix}$. That is, we need to determine if the following vector equation has solution(s)

$$x_1\mathbf{a}_1 + x_2\mathbf{a}_2 = \mathbf{b}, \quad \text{i.e.,} \quad x_1 \begin{bmatrix} 3 \\ 0 \\ 1 \\ 4 \end{bmatrix} + x_2 \begin{bmatrix} 1 \\ -4 \\ 0 \\ 5 \end{bmatrix} = \begin{bmatrix} 3 \\ 12 \\ 2 \\ -7 \end{bmatrix}$$

This is equivalent to the matrix equation

$$A\mathbf{x} = \mathbf{b}, \quad \text{i.e.,} \quad \begin{bmatrix} 3 & 1 \\ 0 & -4 \\ 1 & 0 \\ 4 & 5 \end{bmatrix} \begin{bmatrix} x_1 \\ x_2 \end{bmatrix} = \begin{bmatrix} 3 \\ 12 \\ 2 \\ -7 \end{bmatrix}$$

and also equivalent to the system of linear equations whose augmented matrix is

$$\begin{bmatrix} 3 & 1 & 3 \\ 0 & -4 & 12 \\ 1 & 0 & 2 \\ 4 & 5 & -7 \end{bmatrix}.$$

The **matrix equation**

$$A\mathbf{x} = \mathbf{b}$$

where A is $m \times n$ with its columns $\mathbf{a}_1, \mathbf{a}_2, \ldots, \mathbf{a}_n$, and $\mathbf{b}$ is a vector in R^m, is equivalent to the **vector equation**

$$x_1\mathbf{a}_1 + x_2\mathbf{a}_2 + \cdots + x_n\mathbf{a}_n = \mathbf{b}$$

and also equivalent to the linear system whose augmented matrix is

$$\begin{bmatrix} \mathbf{a}_1 & \mathbf{a}_2 & \cdots & \mathbf{a}_n & \mathbf{b} \end{bmatrix}$$

Example 4 Let A be the 2×4 matrix

$$A = \begin{bmatrix} 1 & 0 & 0 & 0 \\ 0 & 1 & 0 & 0 \end{bmatrix}$$

Let's solve $A\mathbf{x} = \mathbf{b}$ with $\mathbf{b} = \begin{bmatrix} 2 \\ 8 \end{bmatrix}$. Thus, we need to solve

$$\begin{bmatrix} 1 & 0 & 0 & 0 \\ 0 & 1 & 0 & 0 \end{bmatrix} \begin{bmatrix} x_1 \\ x_2 \\ x_3 \\ x_4 \end{bmatrix} = \begin{bmatrix} 2 \\ 8 \end{bmatrix}$$

Clearly, the system has the augmented matrix

$$\begin{bmatrix} 1 & 0 & 0 & 0 & 2 \\ 0 & 1 & 0 & 0 & 8 \end{bmatrix}$$

and the system is consistent and its general solution is

$$\mathbf{x} = \begin{bmatrix} 2 \\ 8 \\ x_3 \\ x_4 \end{bmatrix}$$

where x_3 and x_4 can be any real numbers. This also means that the vector $\mathbf{b} = \begin{bmatrix} 2 \\ 8 \end{bmatrix}$ can be written as linear combination of columns of A as follows,

$$2\begin{bmatrix} 1 \\ 0 \end{bmatrix} + 8\begin{bmatrix} 0 \\ 1 \end{bmatrix} + x_3\begin{bmatrix} 0 \\ 0 \end{bmatrix} + x_4\begin{bmatrix} 0 \\ 0 \end{bmatrix} = \begin{bmatrix} 2 \\ 8 \end{bmatrix}$$

Can each vector $\mathbf{b} = \begin{bmatrix} b_1 \\ b_2 \end{bmatrix}$ in $\mathbf{R}^2$ be represented as a linear combination of columns of A? To answer this question, we need to determine if $A\mathbf{x} = \mathbf{b}$, with $\mathbf{x}$ being a vector in $\mathbf{R}^4$, has a solution. The system is

$$\begin{bmatrix} 1 & 0 & 0 & 0 \\ 0 & 1 & 0 & 0 \end{bmatrix} \begin{bmatrix} x_1 \\ x_2 \\ x_3 \\ x_4 \end{bmatrix} = \begin{bmatrix} b_1 \\ b_2 \end{bmatrix}$$

whose vector equation form is

$$x_1\begin{bmatrix} 1 \\ 0 \end{bmatrix} + x_2\begin{bmatrix} 0 \\ 1 \end{bmatrix} + x_3\begin{bmatrix} 0 \\ 0 \end{bmatrix} + x_4\begin{bmatrix} 0 \\ 0 \end{bmatrix} = \begin{bmatrix} b_1 \\ b_2 \end{bmatrix}$$

with augmented matrix

$$\begin{bmatrix} 1 & 0 & 0 & 0 & b_1 \\ 0 & 1 & 0 & 0 & b_2 \end{bmatrix}$$

We can find the solutions $\mathbf{x} = \begin{bmatrix} b_1 \\ b_2 \\ x_3 \\ x_4 \end{bmatrix}$ for each given $\mathbf{b} = \begin{bmatrix} b_1 \\ b_2 \end{bmatrix}$.

Let A be an $m \times n$ matrix, $\mathbf{x}$ be a vector in $\mathbf{R}^n$, and $\mathbf{b}$ be a vector in $\mathbf{R}^m$. The following statements are equivalent

(a) The equation $A\mathbf{x} = \mathbf{b}$ has a solution for each $\mathbf{b}$ in $\mathbf{R}^m$.

(b) Each $\mathbf{b}$ in $\mathbf{R}^m$ can be written as a linear combination of the columns of A.

(c) The columns of A span $\mathbf{R}^m$.

(d) The matrix A has a pivot position in every row.

1.4 Exercises

Exercise 5 Express each matrix equation in the form of a vector equation.

(a) $$\begin{bmatrix} 3 & -1 & 2 \\ 2 & 0 & 5 \end{bmatrix} \begin{bmatrix} x_1 \\ x_2 \\ x_3 \end{bmatrix} = \begin{bmatrix} 5 \\ 9 \end{bmatrix}$$

(b) $$\begin{bmatrix} 1 & 6 \\ 0 & -4 \\ 5 & 3 \\ 2 & 0 \end{bmatrix} \begin{bmatrix} x_1 \\ x_2 \end{bmatrix} = \begin{bmatrix} 5 \\ 9 \\ 0 \\ -7 \end{bmatrix}$$

Exercise 6 Evaluate each product

(a) $$\begin{bmatrix} 3 & -1 & 2 \\ 2 & 0 & 5 \end{bmatrix} \begin{bmatrix} 1 \\ 3 \\ -2 \end{bmatrix}$$

(b) $$\begin{bmatrix} 1 & 6 \\ 0 & -4 \\ 5 & 3 \\ 2 & 0 \end{bmatrix} \begin{bmatrix} 2 \\ -1 \end{bmatrix}$$

Exercise 7 Express each vector equation in the form of a matrix equation.

(a) $$x_1 \begin{bmatrix} 3 \\ -1 \end{bmatrix} + x_2 \begin{bmatrix} -2 \\ 5 \end{bmatrix} + x_3 \begin{bmatrix} 2 \\ 0 \end{bmatrix} + x_4 \begin{bmatrix} -6 \\ 4 \end{bmatrix} = \begin{bmatrix} 7 \\ 11 \end{bmatrix}$$

$$\text{(b) } x_1 \begin{bmatrix} 2 \\ -1 \\ 4 \end{bmatrix} + x_2 \begin{bmatrix} -1 \\ 4 \\ 6 \end{bmatrix} + x_3 \begin{bmatrix} 3 \\ 1 \\ 8 \end{bmatrix} = \begin{bmatrix} 6 \\ -5 \\ 9 \end{bmatrix}$$

Exercise 8 Express the given linear system first as a matrix equation and then as a vector equation.

$$3x_1 + x_2 - 7x_3 = -9$$
$$6x_1 - 2x_2 + x_3 = 5$$

Exercise 9 Let $\mathbf{u} = \begin{bmatrix} 3 \\ 9 \end{bmatrix}$. Determine if $\mathbf{u}$ is in the subset of $\mathbf{R}^2$ spanned by the columns of $A = \begin{bmatrix} 1 & -2 \\ 2 & -1 \end{bmatrix}$.

Exercise 10 Determine if the vector $\begin{bmatrix} 5 \\ 7 \end{bmatrix}$ is in $Span \left\{ \begin{bmatrix} 3 \\ 0 \end{bmatrix}, \begin{bmatrix} 2 \\ 2 \end{bmatrix} \right\}$.

Exercise 11 Determine if the vector $\begin{bmatrix} 5 \\ 6 \\ -3 \end{bmatrix}$ can be written as a linear combination of the columns of $A = \begin{bmatrix} 1 & 0 & 0 \\ 0 & 1 & 0 \\ 0 & 0 & 1 \end{bmatrix}$.

Exercise 12 Determine if each vector $\mathbf{b} = \begin{bmatrix} b_1 \\ b_2 \\ b_3 \end{bmatrix}$ can be written as a linear combination of the columns of $A = \begin{bmatrix} 1 & 0 & 0 \\ 0 & 1 & 0 \\ 0 & 0 & 1 \end{bmatrix}$. Does the columns of A span $\mathbf{R}^3$?

Exercise 13 Let $\mathbf{u} = \begin{bmatrix} 3 \\ 8 \\ 5 \end{bmatrix}$. Determine if $\mathbf{u}$ is in the subset of $\mathbf{R}^3$ spanned by the columns of $A = \begin{bmatrix} 1 & 3 & 1 & 1 \\ 2 & -2 & 1 & 2 \\ 1 & -5 & 0 & 1 \end{bmatrix}$.

1.5 Homogeneous and Nonhomogeneous Linear Systems

Homogeneous Linear Systems

A linear system is **homogeneous** if it can be written in the form

$$A\mathbf{x} = \mathbf{0}$$

where A is an $m \times n$ matrix, $\mathbf{x}$ is a vector in $\mathbf{R}^n$, and $\mathbf{0}$ is the zero vector in $\mathbf{R}^m$. The system always has a zero solution $\mathbf{x} = \mathbf{0}$ in $\mathbf{R}^n$, which is also called the **trivial solution**. For a specific homogeneous linear system, it is important to know if it has **nontrivial solutions** in addition to the trivial solution.

Example 1 (*Homogeneous Linear Systems*) The following linear system is homogeneous:

$$\begin{aligned} x_1 - 3x_2 - 2x_3 &= 0 \\ x_2 - x_3 &= 0 \end{aligned}$$

In the matrix equation form $A\mathbf{x} = \mathbf{0}$:

$$\begin{bmatrix} 1 & -3 & -2 \\ 0 & 1 & -1 \end{bmatrix} \begin{bmatrix} x_1 \\ x_2 \\ x_3 \end{bmatrix} = \begin{bmatrix} 0 \\ 0 \end{bmatrix}$$

It is obvious that $\mathbf{x} = \begin{bmatrix} 0 \\ 0 \\ 0 \end{bmatrix}$ is a solution of the system. To know if the system has nontrivial solutions, we row reduce the augmented matrix to reduced echelon form:

$$\begin{bmatrix} 1 & -3 & -2 & 0 \\ 0 & 1 & -1 & 0 \end{bmatrix} \sim \begin{bmatrix} 1 & 0 & -5 & 0 \\ 0 & 1 & -1 & 0 \end{bmatrix}$$

That is,

$$x_1 - 5x_3 = 0$$

$$x_2 - x_3 = 0$$

with x_3 being a free variable. The general solution of the system can be expressed as

$$\mathbf{x} = \begin{bmatrix} x_1 \\ x_2 \\ x_3 \end{bmatrix} = \begin{bmatrix} 5x_3 \\ x_3 \\ x_3 \end{bmatrix} = x_3 \begin{bmatrix} 5 \\ 1 \\ 1 \end{bmatrix}$$

Let $\mathbf{u} = \begin{bmatrix} 5 \\ 1 \\ 1 \end{bmatrix}$, then any solution to the system is a scalar multiple of $\mathbf{u}$. That is, $\mathbf{x} = x_3\mathbf{u}$ (x_3 is free), or equivalently $\mathbf{x} = t\mathbf{u}$ (t in $\mathbf{R}$). The trivial solution is obtained when $x_3 = 0$. In other words, the solution set is $Span\ \{\mathbf{u}\}$. Geometrically, the solution set is a line through $\mathbf{0}$ and $\mathbf{u}$ in $\mathbf{R}^3$.

Example 2 (*Homogeneous Linear Systems*) Let's consider another system of equations that is homogeneous:

$$\begin{aligned} x_1 - 3x_2 - 2x_3 &= 0 \\ 2x_1 - 6x_2 - 4x_3 &= 0 \end{aligned}$$

In the matrix equation form $A\mathbf{x} = \mathbf{0}$:

$$\begin{bmatrix} 1 & -3 & -2 \\ 2 & -6 & -4 \end{bmatrix} \begin{bmatrix} x_1 \\ x_2 \\ x_3 \end{bmatrix} = \begin{bmatrix} 0 \\ 0 \end{bmatrix}$$

It is obvious that $\mathbf{x} = \begin{bmatrix} 0 \\ 0 \\ 0 \end{bmatrix}$ is a solution of the system. To know if the system has nontrivial solutions, we row reduce the augmented matrix to reduced echelon form:

$$\begin{bmatrix} 1 & -3 & -2 & 0 \\ 2 & -6 & -4 & 0 \end{bmatrix} \sim \begin{bmatrix} 1 & -3 & -2 & 0 \\ 0 & 0 & 0 & 0 \end{bmatrix}$$

That is,

$$\begin{aligned} x_1 - 3x_2 - 2x_3 &= 0 \\ 0 &= 0 \end{aligned}$$

with two free variables x_2 and x_3. By solving for the basic variable x_1, we obtain the general solution of the system as

$$\mathbf{x} = \begin{bmatrix} x_1 \\ x_2 \\ x_3 \end{bmatrix} = \begin{bmatrix} 3x_2 + 2x_3 \\ x_2 \\ x_3 \end{bmatrix} = \begin{bmatrix} 3x_2 \\ x_2 \\ 0 \end{bmatrix} + \begin{bmatrix} 2x_3 \\ 0 \\ x_3 \end{bmatrix} = x_2 \begin{bmatrix} 3 \\ 1 \\ 0 \end{bmatrix} + x_3 \begin{bmatrix} 2 \\ 0 \\ 1 \end{bmatrix}$$

Let $\mathbf{u} = \begin{bmatrix} 3 \\ 1 \\ 0 \end{bmatrix}$ and $\mathbf{v} = \begin{bmatrix} 2 \\ 0 \\ 1 \end{bmatrix}$. Any solution to the system is a linear combination of $\mathbf{u}$ and $\mathbf{v}$. That is, $\mathbf{x} = x_2\mathbf{u} + x_3\mathbf{v}$ (x_2 and x_3 are free), or equivalently $\mathbf{x} = s\mathbf{u} + t\mathbf{v}$ ($s,\ t$ in $\mathbf{R}$). The trivial solution is obtained when $x_2 = 0$ and $x_3 = 0$. In other words, the solution set is $Span\ \{\mathbf{u},\ \mathbf{v}\}$. Geometrically, the solution set is a plane through $\mathbf{0}$, $\mathbf{u}$, and $\mathbf{v}$ in $\mathbf{R}^3$.

The homogeneous equation $A\mathbf{x} = \mathbf{0}$ has infinitely many nontrivial solutions if and only if it has free variable(s). The equation $A\mathbf{x} = \mathbf{0}$ has only the trivial solution if and only if it does not have any free variable.

Remark 3 The matrix equation $A\mathbf{x} = \mathbf{0}$ has the augmented matrix $\begin{bmatrix} A & \mathbf{0} \end{bmatrix}$. The last column will always remain $\mathbf{0}$ no matter what elementary row operations are applied to the matrix $\begin{bmatrix} A & \mathbf{0} \end{bmatrix}$. For the ease of writing, we can omit the last column $\mathbf{0}$ during the row reduction process.

Nonhomogeneous Linear Systems

A linear system is **nonhomogeneous** if it can be written in the form

$$A\mathbf{x} = \mathbf{b}$$

where A is an $m \times n$ matrix, $\mathbf{x}$ is a vector in $\mathbf{R}^n$, and $\mathbf{b}$ is a nonzero vector in $\mathbf{R}^m$. Note that it is impossible that a nonhomogeneous system has a zero solution $\mathbf{x} = \mathbf{0}$ in $\mathbf{R}^n$.

Example 4 (*Nonhomogeneous Linear Systems*) The following linear system is nonhomogeneous:

$$\begin{aligned} x_1 - 3x_2 - 2x_3 &= 1 \\ x_2 - x_3 &= 2 \end{aligned}$$

The augmented matrix is reduced to the reduced echelon form:

$$\begin{bmatrix} 1 & -3 & -2 & 1 \\ 0 & 1 & -1 & 2 \end{bmatrix} \sim \begin{bmatrix} 1 & 0 & -5 & 7 \\ 0 & 1 & -1 & 2 \end{bmatrix}$$

That is,

$$\begin{aligned} x_1 - 5x_3 &= 7 \\ x_2 - x_3 &= 2 \end{aligned}$$

with x_3 being a free variable. The general solution of the system is

$$\mathbf{x} = \begin{bmatrix} x_1 \\ x_2 \\ x_3 \end{bmatrix} = \begin{bmatrix} 7 + 5x_3 \\ 2 + x_3 \\ x_3 \end{bmatrix} = \begin{bmatrix} 7 \\ 2 \\ 0 \end{bmatrix} + \begin{bmatrix} 5x_3 \\ x_3 \\ x_3 \end{bmatrix} = \begin{bmatrix} 7 \\ 2 \\ 0 \end{bmatrix} + x_3 \begin{bmatrix} 5 \\ 1 \\ 1 \end{bmatrix}$$

Let $\mathbf{p} = \begin{bmatrix} 7 \\ 2 \\ 0 \end{bmatrix}$ and $\mathbf{u} = \begin{bmatrix} 5 \\ 1 \\ 1 \end{bmatrix}$, the general solution of the system can be represented as $\mathbf{x} = \mathbf{p} + x_3\mathbf{u}$.

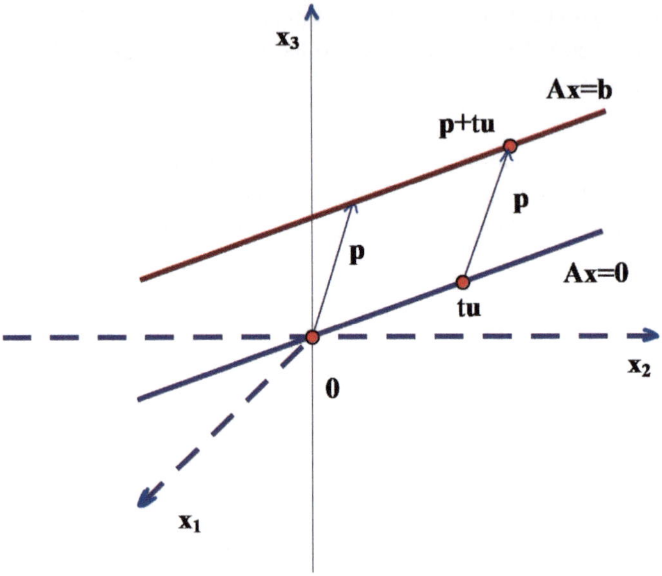

Fig. 2 Homogeneous and nonhomogeneous solutions, with 1 free variable

Remark 5 We have solved

$$A\mathbf{x} = \mathbf{b}, \text{ i.e., } \begin{bmatrix} 1 & -3 & -2 \\ 0 & 1 & -1 \end{bmatrix} \begin{bmatrix} x_1 \\ x_2 \\ x_3 \end{bmatrix} = \begin{bmatrix} 1 \\ 2 \end{bmatrix}$$

and its corresponding homogeneous equation

$$A\mathbf{x} = \mathbf{0}, \text{ i.e., } \begin{bmatrix} 1 & -3 & -2 \\ 0 & 1 & -1 \end{bmatrix} \begin{bmatrix} x_1 \\ x_2 \\ x_3 \end{bmatrix} = \begin{bmatrix} 0 \\ 0 \end{bmatrix}$$

The solution set of the nonhomogeneous equation consists all vectors $\mathbf{p}+t\mathbf{u}$ (t in $\mathbf{R}$). The solution set of the corresponding homogeneous equation consists all vectors $t\mathbf{u}$ (t in $\mathbf{R}$). The vector $\mathbf{p}$ is just a particular solution of $A\mathbf{x} = \mathbf{b}$. But this one solution of $A\mathbf{x} = \mathbf{b}$ allows us to find all solutions of $A\mathbf{x} = \mathbf{b}$ as long as we have solved $A\mathbf{x} = \mathbf{0}$. By adding $\mathbf{p}$ to the solutions of $A\mathbf{x} = \mathbf{0}$, we obtain the solutions of $A\mathbf{x} = \mathbf{b}$. See Fig. 2 for an illustration of the relationship between solutions of homogeneous and nonhomogeneous equations. The solution sets of $A\mathbf{x} = \mathbf{0}$ and $A\mathbf{x} = \mathbf{b}$ are parallel lines.

Example 6 (*Nonhomogeneous Linear Systems*) Consider the following nonhomogeneous equation, for which the corresponding homogeneous equation is already solved:

$$\begin{bmatrix} 1 & -3 & -2 \\ 2 & -6 & -4 \end{bmatrix} \begin{bmatrix} x_1 \\ x_2 \\ x_3 \end{bmatrix} = \begin{bmatrix} 1 \\ 2 \end{bmatrix}$$

We row reduce the augmented matrix to its reduced echelon form:

$$\begin{bmatrix} 1 & -3 & -2 & 1 \\ 2 & -6 & -4 & 2 \end{bmatrix} \sim \begin{bmatrix} 1 & -3 & -2 & 1 \\ 0 & 0 & 0 & 0 \end{bmatrix}$$

That is,

$$\begin{aligned} x_1 - 3x_2 - 2x_3 &= 1 \\ 0 &= 0 \end{aligned}$$

with two free variables x_2 and x_3. By solving for the basic variable x_1, we obtain the general solution of the system as

$$\mathbf{x} = \begin{bmatrix} x_1 \\ x_2 \\ x_3 \end{bmatrix} = \begin{bmatrix} 1 + 3x_2 + 2x_3 \\ x_2 \\ x_3 \end{bmatrix} = \begin{bmatrix} 1 \\ 0 \\ 0 \end{bmatrix} + \begin{bmatrix} 3x_2 \\ x_2 \\ 0 \end{bmatrix} + \begin{bmatrix} 2x_3 \\ 0 \\ x_3 \end{bmatrix} = \begin{bmatrix} 1 \\ 0 \\ 0 \end{bmatrix} + x_2 \begin{bmatrix} 3 \\ 1 \\ 0 \end{bmatrix} + x_3 \begin{bmatrix} 2 \\ 0 \\ 1 \end{bmatrix}$$

Let $\mathbf{p} = \begin{bmatrix} 1 \\ 0 \\ 0 \end{bmatrix}$, $\mathbf{u} = \begin{bmatrix} 3 \\ 1 \\ 0 \end{bmatrix}$, and $\mathbf{v} = \begin{bmatrix} 2 \\ 0 \\ 1 \end{bmatrix}$. Any solution $\mathbf{x}$ to the system is in the form of $\mathbf{x} = \mathbf{p} + x_2\mathbf{u} + x_3\mathbf{v}$ (x_2 and x_3 are free), or equivalently $\mathbf{x} = \mathbf{p} + s\mathbf{u} + t\mathbf{v}$ (s, t in $\mathbf{R}$).

Remark 7 We have solved both the equation

$$A\mathbf{x} = \mathbf{b}, \text{ i.e., } \begin{bmatrix} 1 & -3 & -2 \\ 2 & -6 & -4 \end{bmatrix} \begin{bmatrix} x_1 \\ x_2 \\ x_3 \end{bmatrix} = \begin{bmatrix} 1 \\ 2 \end{bmatrix}$$

and its corresponding homogeneous equation

$$A\mathbf{x} = \mathbf{0}, \text{ i.e., } \begin{bmatrix} 1 & -3 & -2 \\ 2 & -6 & -4 \end{bmatrix} \begin{bmatrix} x_1 \\ x_2 \\ x_3 \end{bmatrix} = \begin{bmatrix} 0 \\ 0 \end{bmatrix}$$

The general solution for $A\mathbf{x} = \mathbf{b}$ is $\mathbf{x} = \mathbf{p} + x_2\mathbf{u} + x_3\mathbf{v}$. Note that $\mathbf{p}$ is a particular solution with $x_2 = 0$ and $x_3 = 0$. A particular solution $\mathbf{p}$ can lead to the finding of all solutions of $A\mathbf{x} = \mathbf{b}$, as long as we know the general solution $x_2\mathbf{u} + x_3\mathbf{v}$ of $A\mathbf{x} = \mathbf{0}$. Geometrically, we know the solution set of $A\mathbf{x} = \mathbf{0}$ is a plane through $\mathbf{0}$, $\mathbf{u}$, and $\mathbf{v}$ in $\mathbf{R}^3$. The solution set of $A\mathbf{x} = \mathbf{b}$ must

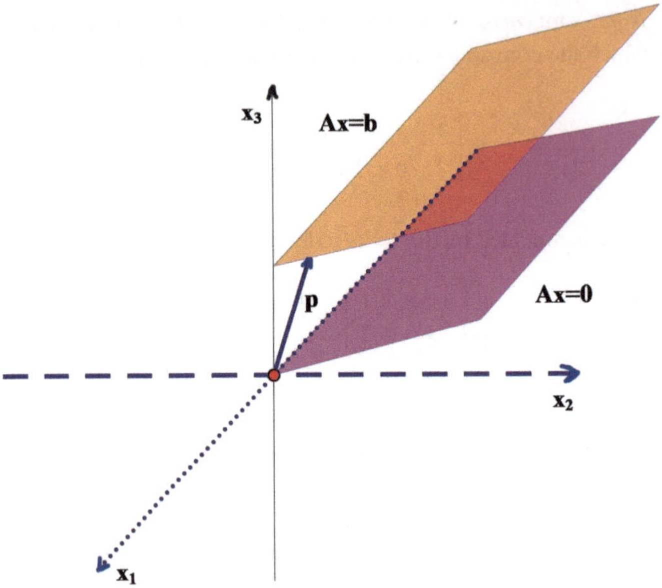

Fig. 3 Homogeneous and nonhomogeneous solutions, with 2 free variables

be a plane in $\mathbf{R}^3$ not passing through $\mathbf{0}$, and this plane is obtained by translating the solution set of $A\mathbf{x} = \mathbf{0}$ using any particular solution $\mathbf{p}$ of $A\mathbf{x} = \mathbf{b}$. See Fig. 3 for an illustration of the relationship between solutions of homogeneous and nonhomogeneous equations.

1.5 Exercises

Exercise 8 Find the solution set for each homogeneous system

a.
$$\begin{aligned} x_1 \quad -3x_2 -2x_3 &= 0 \\ x_2 \quad -x_3 &= 0 \\ -2x_1 +3x_2 +7x_3 &= 0 \end{aligned}$$

b. $$2x_1 - 6x_2 + 3x_3 = 0$$

c. $$x_1 - 5x_2 - x_3 + 2x_4 = 0$$

Exercise 9 Find the solution set for each nonhomogeneous system

a.
$$\begin{aligned} x_1 \quad -3x_2 -2x_3 &= -5 \\ x_2 \quad -x_3 &= 4 \\ -2x_1 +3x_2 +7x_3 &= -2 \end{aligned}$$

b. $2x_1 - 6x_2 + 3x_3 = 8$

c. $x_1 - 5x_2 - x_3 + 2x_4 = 1$

Exercise 10 Determine if the vector $\mathbf{0} = \begin{bmatrix} 0 \\ 0 \\ 0 \end{bmatrix}$ is a linear combination of the columns of $A = \begin{bmatrix} 1 & -3 \\ 0 & 1 \\ -2 & 3 \end{bmatrix}$.

Exercise 11 Determine if the vector $\mathbf{b} = \begin{bmatrix} 2 \\ 1 \\ 7 \end{bmatrix}$ is a linear combination of the columns of $A = \begin{bmatrix} 1 & -1 & 2 & 3 \\ 2 & 1 & 1 & 0 \\ 1 & 2 & -1 & -3 \end{bmatrix}$.

Exercise 12 Find the solution set for each homogeneous equation $A\mathbf{x} = \mathbf{0}$ with given A.

a. $A = \begin{bmatrix} 1 & 1 & 0 \\ 0 & 0 & 5 \end{bmatrix}$

b. $A = \begin{bmatrix} 1 & 1 & 0 \\ 1 & 2 & 1 \\ 0 & 1 & 2 \end{bmatrix}$

c. $A = \begin{bmatrix} 1 & 0 & -3 & 0 & 2 \\ 0 & 1 & 1 & 0 & 1 \\ 0 & 0 & 0 & 1 & 6 \end{bmatrix}$

Exercise 13 Find the general solution of each equation $A\mathbf{x} = \mathbf{b}$.

a. $A = \begin{bmatrix} 1 & 1 & 0 \\ 0 & 0 & 5 \end{bmatrix}, \quad \mathbf{b} = \begin{bmatrix} 3 \\ -10 \end{bmatrix}$

b. $A = \begin{bmatrix} 1 & 1 & 0 \\ 1 & 2 & 1 \\ 0 & 1 & 2 \end{bmatrix}, \quad \mathbf{b} = \begin{bmatrix} 2 \\ 0 \\ -2 \end{bmatrix}$

c. $A = \begin{bmatrix} 1 & 0 & -3 & 0 & 2 \\ 0 & 1 & 1 & 0 & 1 \\ 0 & 0 & 0 & 1 & 6 \end{bmatrix}, \quad \mathbf{b} = \begin{bmatrix} 3 \\ -5 \\ 8 \end{bmatrix}$

1.6 Applications of Linear Systems

Curve Fitting

The curve fitting application involves the use of a certain type of curve to fit a given set of data points.

Example 1 (*Curve Fitting*) Is there a line passing through the points (1, 7), (2, 9), and (−3, − 1)? We know that two different points can determine a line. But we don't know if we can find a line that will pass through all three points. Recall that a line in $\mathbf{R}^2$ has an equation of the general form

$$ax + by = c$$

The three points are on the line if and only if all three points satisfy the equation. Then we have a linear system:

$$\begin{aligned} a + 7b &= c \\ 2a + 9b &= c \\ -3a - b &= c \end{aligned}$$

which is a homogeneous system:

$$\begin{aligned} a + 7b - c &= 0 \\ 2a + 9b - c &= 0 \\ -3a - b - c &= 0 \end{aligned}$$

Its augmented matrix is

$$\begin{bmatrix} 1 & 7 & -1 & 0 \\ 2 & 9 & -1 & 0 \\ -3 & -1 & -1 & 0 \end{bmatrix} \sim \begin{bmatrix} 1 & 7 & -1 & 0 \\ 0 & -5 & 1 & 0 \\ 0 & 20 & -4 & 0 \end{bmatrix} \sim \begin{bmatrix} 1 & 7 & -1 & 0 \\ 0 & -5 & 1 & 0 \\ 0 & 0 & 0 & 0 \end{bmatrix} \sim$$

$$\begin{bmatrix} 1 & 7 & -1 & 0 \\ 0 & 1 & -\frac{1}{5} & 0 \\ 0 & 0 & 0 & 0 \end{bmatrix} \sim \begin{bmatrix} 1 & 0 & \frac{2}{5} & 0 \\ 0 & 1 & -\frac{1}{5} & 0 \\ 0 & 0 & 0 & 0 \end{bmatrix}$$

The reduced echelon form leads to the general solution

$$\begin{aligned} a &= -\frac{2}{5}c \\ b &= \frac{1}{5}c \end{aligned}$$

where c is free.

Therefore, we can obtain the equation of a line with $c \neq 0$ as follows

$$\left(-\frac{2}{5}c\right)x + \left(\frac{1}{5}c\right)y = c$$

By simplifying,

$$2x - y = -5 \quad \text{i.e.,} \quad y = 2x + 5$$

Note that the three given points can not possibly be on a vertical line due to the fact that their x-coordinates are not the same. We can also determine if there is a line in the form of $y = mx + b$ passing through the points $(1,\ 7)$, $(2,\ 9)$, and $(-3,\ -1)$. Then we want to know if there exists a solution for the nonhomogeneous system:

$$\begin{aligned} m + b &= 7 \\ 2m + b &= 9 \\ -3m + b &= -1 \end{aligned}$$

with the augmented matrix

$$\begin{bmatrix} 1 & 1 & 7 \\ 2 & 1 & 9 \\ -3 & 1 & -1 \end{bmatrix} \sim \begin{bmatrix} 1 & 1 & 7 \\ 0 & -1 & -5 \\ 0 & 4 & 20 \end{bmatrix} \sim \begin{bmatrix} 1 & 0 & 2 \\ 0 & 1 & 5 \\ 0 & 0 & 0 \end{bmatrix}$$

We thus obtain $m = 2$ and $b = 5$. The line in slope-intercept form is $y = 2x + 5$.

Example 2 (*Curve Fitting*) Determine if there is a parabola $y = ax^2 + bx + c$ that passes through the points $(1,\ 2)$, $(-2,\ 11)$, and $(3,\ 6)$. The three points are on the parabola if and only if they all satisfy the equation $y = ax^2 + bx + c$. The resulted linear system is

$$\begin{aligned} a\,(1)^2 + b\,(1) + c &= 2 \\ a\,(-2)^2 + b\,(-2) + c &= 11 \\ a\,(3)^2 + b\,(3) + c &= 6 \end{aligned}$$

or in the form of matrix equation:

$$\begin{bmatrix} 1 & 1 & 1 \\ 4 & -2 & 1 \\ 9 & 3 & 1 \end{bmatrix} \begin{bmatrix} a \\ b \\ c \end{bmatrix} = \begin{bmatrix} 2 \\ 11 \\ 6 \end{bmatrix}$$

We reduce its augmented matrix

$$\begin{bmatrix} 1 & 1 & 1 & 2 \\ 4 & -2 & 1 & 11 \\ 9 & 3 & 1 & 6 \end{bmatrix} \sim \begin{bmatrix} 1 & 1 & 1 & 2 \\ 0 & -6 & -3 & 3 \\ 0 & -6 & -8 & -12 \end{bmatrix} \sim \begin{bmatrix} 1 & 1 & 1 & 2 \\ 0 & -6 & -3 & 3 \\ 0 & 0 & -5 & -15 \end{bmatrix} \sim$$

$$\begin{bmatrix} 1 & 0 & \frac{1}{2} & 2\frac{1}{2} \\ 0 & 1 & \frac{1}{2} & -\frac{1}{2} \\ 0 & 0 & -5 & -15 \end{bmatrix} \sim \begin{bmatrix} 1 & 0 & \frac{1}{2} & 2\frac{1}{2} \\ 0 & 1 & \frac{1}{2} & -\frac{1}{2} \\ 0 & 0 & 1 & 3 \end{bmatrix} \sim \begin{bmatrix} 1 & 0 & 0 & 1 \\ 0 & 1 & 0 & -2 \\ 0 & 0 & 1 & 3 \end{bmatrix}$$

There is a unique solution $\begin{bmatrix} a \\ b \\ c \end{bmatrix} = \begin{bmatrix} 1 \\ -2 \\ 3 \end{bmatrix}$ for the above linear system. So there is a unique parabola $y = x^2 - 2x + 3$.

Chemical Reactions

There is a fundamental mathematical principle for chemical equations. That is, we need to balance the numbers of atoms of substances before and after a chemical reaction.

Example 3 Let's consider the chemical reaction

$$(x_1)\,H_2O + (x_2)\,Fe \leftrightarrows (x_3)\,F_e\,(OH)_3 + (x_4)\,H_2$$

To balance this equation, we need to determine x_1, x_2, x_3, and x_4, i.e., the quantities of H_2O, Fe, $F_e\,(OH)_3$, and H_2, so that the numbers of H, O, and Fe atoms are balanced. The number of the H atom should be equal for both sides of the equation,

$$2x_1 = 3x_3 + 2x_4$$

so are the number of the O atom

$$x_1 = 3x_3$$

and that of the Fe atom

$$x_2 = x_3$$

We obtain the following homogeneous system by moving all the variables to the left side

$$\begin{aligned} 2x_1 \quad\quad -3x_3 - 2x_4 &= 0 \\ x_1 \quad\quad -3x_3 \quad\quad &= 0 \\ x_2 - x_3 \quad\quad &= 0 \end{aligned}$$

whose augmented matrix is reduced:

$$\begin{bmatrix} 2 & 0 & -3 & -2 & 0 \\ 1 & 0 & -3 & 0 & 0 \\ 0 & 1 & -1 & 0 & 0 \end{bmatrix} \sim \begin{bmatrix} 1 & 0 & -3 & 0 & 0 \\ 2 & 0 & -3 & -2 & 0 \\ 0 & 1 & -1 & 0 & 0 \end{bmatrix} \sim \begin{bmatrix} 1 & 0 & -3 & 0 & 0 \\ 0 & 0 & 3 & -2 & 0 \\ 0 & 1 & -1 & 0 & 0 \end{bmatrix} \sim$$

$$\begin{bmatrix} 1 & 0 & -3 & 0 & 0 \\ 0 & 1 & -1 & 0 & 0 \\ 0 & 0 & 3 & -2 & 0 \end{bmatrix} \sim \begin{bmatrix} 1 & 0 & 0 & -2 & 0 \\ 0 & 1 & 0 & -\frac{2}{3} & 0 \\ 0 & 0 & 1 & -\frac{2}{3} & 0 \end{bmatrix}$$

The basic variables are x_1, x_2, and x_3. The variable x_4 is free. The linear system has a general solution

$$\begin{bmatrix} x_1 \\ x_2 \\ x_3 \\ x_4 \end{bmatrix} = x_4 \begin{bmatrix} 2 \\ \frac{2}{3} \\ \frac{2}{3} \\ 1 \end{bmatrix}$$

or in the form of parametric vector equation: $\begin{bmatrix} x_1 \\ x_2 \\ x_3 \\ x_4 \end{bmatrix} = t \begin{bmatrix} 2 \\ \frac{2}{3} \\ \frac{2}{3} \\ 1 \end{bmatrix}$, t is any real number. As we know, if t can be any real number, then $3t$ can be any real number. It will not hurt if we expression the general solution in the form $\begin{bmatrix} x_1 \\ x_2 \\ x_3 \\ x_4 \end{bmatrix} = (3t) \begin{bmatrix} 2 \\ \frac{2}{3} \\ \frac{2}{3} \\ 1 \end{bmatrix} = t \begin{bmatrix} 6 \\ 2 \\ 2 \\ 3 \end{bmatrix}$. Substituting the values of x_1, x_2, x_3, and x_4 into the reaction equation,

$$(x_1)\, H_2O + (x_2)\, Fe \leftrightarrows (x_3)\, F_e\, (OH)_3 + (x_4)\, H_2$$

to have

$$(6t)\, H_2O + (2t)\, Fe \leftrightarrows (2t)\, F_e\, (OH)_3 + (3t)\, H_2$$

which implies that the found quantities of substances H_2O, Fe, $F_e\,(OH)_3$, and H_2 will balance the chemical reaction. For the reaction equation, we can also employ a vector equation to describe the balance. Each molecule of H_2O has respectively 2, 1, and 0 atoms of H, O, Fe; each molecule of Fe has respectively 0, 0, and 1 atoms of H, O, Fe; each molecule of $F_e\,(OH)_3$ has respectively 3, 3, and 1 atoms of H, O, Fe; and each molecule of H_2 has respectively 2, 0, and 0 atoms of H, O, Fe; Thus we can describe the reaction equation using the vector equation:

$$x_1 \begin{bmatrix} 2 \\ 1 \\ 0 \end{bmatrix} + x_2 \begin{bmatrix} 0 \\ 0 \\ 1 \end{bmatrix} = x_3 \begin{bmatrix} 3 \\ 3 \\ 1 \end{bmatrix} + x_4 \begin{bmatrix} 2 \\ 0 \\ 0 \end{bmatrix}$$

which is equivalent the homogeneous equation

$$\begin{bmatrix} 2 & 0 & -3 & -2 \\ 1 & 0 & -3 & 0 \\ 0 & 1 & -1 & 0 \end{bmatrix} \begin{bmatrix} x_1 \\ x_2 \\ x_3 \\ x_4 \end{bmatrix} = \begin{bmatrix} 0 \\ 0 \\ 0 \end{bmatrix}$$

Obviously we can find its general solution as $\begin{bmatrix} x_1 \\ x_2 \\ x_3 \\ x_4 \end{bmatrix} = t \begin{bmatrix} 6 \\ 2 \\ 2 \\ 3 \end{bmatrix}$.

1.6 Exercises

Exercise 4 Determine if there exists a line that passes through the points $(1,\ 4)$, $(2,\ 9)$, and $(-2,\ -6)$. Find it if there is one.

Exercise 5 Find a parabola $y = ax^2 + bx + c$ that passes through the three points $(1,\ 4)$, $(-1,\ 6)$ and $(0,\ 3)$.

Exercise 6 Determine if there exists a parabola $y = ax^2 + bx + c$ that passes through the four points $(1,\ 4)$, $(-1,\ 6)$, $(0,\ 3)$ and $(2,\ 6)$. Find it if there is one.

Exercise 7 There are 4 atoms B (boron), S (sulfur), H (hydrogen), and O (oxygen) that are involved in the chemical reaction $B_2S_3 + H_2O \rightarrow H_3BO_3 + H_2S$. Determine the quantities of the substances B_2S_3, H_2O, H_3BO_3, and H_2S to make it a balanced equation.

1.7 Linear Independence

Linear Independent Set of Vectors

The set of n vectors $\{\mathbf{v}_1,\ \mathbf{v}_2,\ \ldots,\ \mathbf{v}_n\}$ are said to be **linearly dependent** if there is at least one vector in the set that can be written as a linear combination of the rest of the vectors. Let's suppose $\mathbf{v}_n$ is a linear combination of the rest of the vectors, that is, we can find values for $c_1, c_2, \ldots, c_{n-1}$ such that

$$\mathbf{v}_n = c_1\mathbf{v}_1 + c_2\mathbf{v}_2 + \cdots + c_{n-1}\mathbf{v}_{n-1}$$

which is equivalent to saying that the homogeneous system

$$c_1\mathbf{v}_1 + c_2\mathbf{v}_2 + \cdots + c_{n-1}\mathbf{v}_{n-1} + c_n\mathbf{v}_n = \mathbf{0}$$

has at least a nontrivial solution $(c_1,\ c_2,\ \ldots,\ c_n)$ with $c_n \neq 0$. On the other hand, if there exists a nontrivial solution to the system $c_1\mathbf{v}_1 + c_2\mathbf{v}_2 + \cdots + c_{n-1}\mathbf{v}_{n-1} + c_n\mathbf{v}_n = \mathbf{0}$, say $(c_1,\ c_2,\ \ldots,\ c_n)$ with $c_i \neq 0$ being the first weight among all weights that is non zero, we would know that $\mathbf{v}_i$ can be written as a linear combination of the rest of the vectors of the set. This is because $c_i \neq 0$ implying

$$-c_i\mathbf{v}_i = c_1\mathbf{v}_1 + c_2\mathbf{v}_2 + \cdots + c_{i-1}\mathbf{v}_{i-1} + c_{i+1}\mathbf{v}_{i+1} + \cdots + c_n\mathbf{v}_n$$

that is,

$$\mathbf{v}_i = \left(-\frac{c_1}{c_i}\right)\mathbf{v}_1 + \left(-\frac{c_2}{c_i}\right)\mathbf{v}_2 + \cdots + \left(-\frac{c_{i-1}}{c_i}\right)\mathbf{v}_{i-1} + \left(-\frac{c_{i+1}}{c_i}\right)\mathbf{v}_{i+1} + \cdots + \left(-\frac{c_n}{c_i}\right)\mathbf{v}_n$$

The set of n vectors $\{\mathbf{v}_1, \ \mathbf{v}_2, \ \ldots, \ \mathbf{v}_n\}$ is said to be **linearly independent** if the vector equation

$$c_1\mathbf{v}_1 + c_2\mathbf{v}_2 + \cdots + c_{n-1}\mathbf{v}_{n-1} + c_n\mathbf{v}_n = \mathbf{0}$$

has only the trivial solution $(c_1, \ c_2, \ \ldots, \ c_n) = (0, \ 0, \ \ldots, 0)$.

Example 1 (*Two Vectors*) Determine if the set of two vectors $\{\mathbf{v}_1, \ \mathbf{v}_2\}$, with $\mathbf{v}_1 = \begin{pmatrix} -1 \\ 2 \end{pmatrix}$ and $\mathbf{v}_2 = \begin{pmatrix} -3 \\ 6 \end{pmatrix}$, is linearly independent. Let's consider the vector equation

$$c_1\mathbf{v}_1 + c_2\mathbf{v}_2 = \mathbf{0}$$

i.e.,

$$c_1\begin{pmatrix} -1 \\ 2 \end{pmatrix} + c_2\begin{pmatrix} -3 \\ 6 \end{pmatrix} = \begin{pmatrix} 0 \\ 0 \end{pmatrix}.$$

The augmented matrix of the system

$$\begin{pmatrix} -1 & -3 & 0 \\ 2 & 6 & 0 \end{pmatrix}$$

can be row reduced to

$$\begin{pmatrix} -1 & -3 & 0 \\ 0 & 0 & 0 \end{pmatrix}$$

which implies nontrivial solutions to the system. Therefore, the set of two vectors $\{\mathbf{v}_1, \ \mathbf{v}_2\}$ is not linearly independent. As a matter of fact, suppose we pick a nontrivial solution, say $(c_1, \ c_2) = (-3, \ 1)$, then

$$-3\mathbf{v}_1 + \mathbf{v}_2 = \mathbf{0}$$

which means that one vector can be represented as the multiple of the other.

Remark 2 The set of two vectors is relatively easy for us to determine its linear independence. In the previous example, if we notice $\mathbf{v}_2 = 3\mathbf{v}_1$, we know that the set is linearly dependent. A nontrivial solution exists for $c_1\mathbf{v}_1 + c_2\mathbf{v}_2 = \mathbf{0}$. What if we have a set of only two vectors $\{\mathbf{v}_1, \ \mathbf{v}_2\}$ with $\mathbf{v}_1$ and $\mathbf{v}_2$ are not multiples of one another? The system

$$c_1\mathbf{v}_1 + c_2\mathbf{v}_2 = \mathbf{0}$$

must have only the trivial solution $c_1 = 0$ and $c_2 = 0$, implying the linear independence. Otherwise if $c_1 \neq 0$, the $\mathbf{v}_1$ is a multiple of $\mathbf{v}_2$, or if $c_2 \neq 0$, the $\mathbf{v}_2$ is a multiple of $\mathbf{v}_1$. Either case will contradict the assumption that $\mathbf{v}_1$ and $\mathbf{v}_2$ are not multiples of one another.

Example 3 (*Two Vectors*) Determine if the set of two vectors $\{\mathbf{v}_1, \ \mathbf{v}_2\}$, with $\mathbf{v}_1 = \begin{pmatrix} -1 \\ 2 \\ 0 \end{pmatrix}$ and $\mathbf{v}_2 = \begin{pmatrix} 4 \\ -8 \\ 0 \end{pmatrix}$, is linearly independent. We observe that $\mathbf{v}_2 = -4\mathbf{v}_1$. This is a linearly dependent set. The alternative way is to consider $c_1\mathbf{v}_1 + c_2\mathbf{v}_2 = \mathbf{0}$, with its augmented matrix

$$\begin{pmatrix} -1 & 4 & 0 \\ 2 & -8 & 0 \\ 0 & 0 & 0 \end{pmatrix}$$

being row reduced to

$$\begin{pmatrix} -1 & 4 & 0 \\ 0 & 0 & 0 \\ 0 & 0 & 0 \end{pmatrix}$$

we know there is at least a nontrivial solution (c_1, c_2), implying the dependence of the two vectors.

Example 4 (*Two Vectors*) Determine if the set of two vectors $\{\mathbf{v}_1, \ \mathbf{v}_2\}$, with $\mathbf{v}_1 = \begin{pmatrix} -1 \\ 2 \\ 0 \end{pmatrix}$ and $\mathbf{v}_2 = \begin{pmatrix} 4 \\ -8 \\ 1 \end{pmatrix}$, is linearly independent. Since $\mathbf{v}_1$ and $\mathbf{v}_2$ are not multiples of one another, they are linearly independent. The alternative way is to consider $c_1\mathbf{v}_1 + c_2\mathbf{v}_2 = \mathbf{0}$, with its augmented matrix

$$\begin{pmatrix} -1 & 4 & 0 \\ 2 & -8 & 0 \\ 0 & 1 & 0 \end{pmatrix}$$

being row reduced to

$$\begin{pmatrix} -1 & 4 & 0 \\ 0 & 0 & 0 \\ 0 & 1 & 0 \end{pmatrix}$$

and further reduced to

$$\begin{pmatrix} 1 & 0 & 0 \\ 0 & 1 & 0 \\ 0 & 0 & 0 \end{pmatrix}$$

indicating only a trivial solution exists.

Example 5 (*Three Vectors*) Determine if the set of three vectors $\{\mathbf{v}_1, \ \mathbf{v}_2, \ \mathbf{v}_3\}$ is linearly independent. Here $\mathbf{v}_1 = \begin{pmatrix} 3 \\ 1 \end{pmatrix}$, $\mathbf{v}_2 = \begin{pmatrix} 6 \\ 1 \end{pmatrix}$, and $\mathbf{v}_3 = \begin{pmatrix} 0 \\ 0 \end{pmatrix}$. To determine the relation among $\mathbf{v}_1$, $\mathbf{v}_2$, and $\mathbf{v}_3$, we consider the homogeneous system

$$x_1\mathbf{v}_1 + x_2\mathbf{v}_2 + x_3\mathbf{v}_3 = \mathbf{0}$$

i.e.,

$$x_1 \begin{pmatrix} 3 \\ 1 \end{pmatrix} + x_2 \begin{pmatrix} 6 \\ 1 \end{pmatrix} + x_3 \begin{pmatrix} 0 \\ 0 \end{pmatrix} = \begin{pmatrix} 0 \\ 0 \end{pmatrix}$$

which is the same as

$$\begin{aligned} 3x_1 + 6x_2 + 0x_3 &= 0 \\ x_1 + x_2 + 0x_3 &= 0 \end{aligned}$$

We must have at least a free variable since the number of variables is more than the number of equations. Therefore, the system has nontrivial solutions and the set of vectors is dependent. Note that there is a zero vector $\mathbf{v}_3 = \begin{pmatrix} 0 \\ 0 \end{pmatrix}$ in this particular set, and it can be easily represented as a linear combination of $\mathbf{v}_1$ and $\mathbf{v}_2$, i.e., $\mathbf{v}_3 = 0\mathbf{v}_1 + 0\mathbf{v}_2$. This means a nontrivial solution such as $(x_1, x_2, x_3) = (0, 0, 1)$ exists for the system $x_1\mathbf{v}_1 + x_2\mathbf{v}_2 + x_3\mathbf{v}_3 = \mathbf{0}$.

Remark 6 The set of vectors consisting a zero vector must be a linearly dependent set.

Matrix Columns and Their Linear Independence

A matrix A with n columns can be expressed as $A = [\mathbf{a}_1, \ \mathbf{a}_2, \ \ldots, \ \mathbf{a}_n]$. The columns of A are linearly independent if and only if the vector equation

$$x_1\mathbf{a}_1 + x_2\mathbf{a}_2 + \cdots + x_n\mathbf{a}_n = \mathbf{0}$$

has only the trivial solution. That is, the columns of A are linearly independent if and only if the matrix equation

$$[\mathbf{a}_1, \quad \mathbf{a}_2, \quad \ldots, \quad \mathbf{a}_n] \begin{bmatrix} x_1 \\ x_2 \\ \vdots \\ x_n \end{bmatrix} = \mathbf{0}$$

has only the trivial solution.

Example 7 (*Matrix Columns*) Determine if the columns of

$$A = \begin{bmatrix} 1 & -3 & -2 \\ 0 & 1 & -1 \\ -2 & 3 & 7 \end{bmatrix}$$

form a linearly independent set. The vector equation

$$x_1 \begin{bmatrix} 1 \\ 0 \\ -2 \end{bmatrix} + x_2 \begin{bmatrix} -3 \\ 1 \\ 3 \end{bmatrix} + x_3 \begin{bmatrix} -2 \\ -1 \\ 7 \end{bmatrix} = \begin{bmatrix} 0 \\ 0 \\ 0 \end{bmatrix}$$

has an augmented matrix that can be reduced to

$$\begin{bmatrix} 1 & 0 & -5 & 0 \\ 0 & 1 & -1 & 0 \\ 0 & 0 & 0 & 0 \end{bmatrix}$$

With x_3 being a free variable, the system has nontrivial solutions

$$\begin{bmatrix} x_1 \\ x_2 \\ x_3 \end{bmatrix} = x_3 \begin{bmatrix} 5 \\ 1 \\ 1 \end{bmatrix}$$

Therefore the columns of A are linearly dependent.

Example 8 (*Matrix Columns*) Determine if the columns of

$$A = \begin{bmatrix} 1 & -3 & -2 \\ 0 & 1 & -1 \end{bmatrix}$$

form a linearly independent set. The equation $A\mathbf{x} = \mathbf{0}$

$$\begin{bmatrix} 1 & -3 & -2 \\ 0 & 1 & -1 \end{bmatrix} \begin{bmatrix} x_1 \\ x_2 \\ x_3 \end{bmatrix} = \begin{bmatrix} 0 \\ 0 \\ 0 \end{bmatrix}$$

is the same as the system

$$x_1 - 3x_2 - 2x_3 = 0$$
$$x_2 - x_3 = 0$$

Since there are two equations and three variables, we must have at least one free variable. There exists a nontrivial solution $\mathbf{x}$. So the columns of A are linearly dependent.

Example 9 (*Matrix Columns*) Determine if the columns of

$$A = \begin{bmatrix} 1 & 3 & 6 & 1 & 2 \\ 5 & 2 & 7 & 1 & 4 \\ 3 & 1 & 3 & 1 & 8 \end{bmatrix}$$

form a linearly independent set. The homogeneous system $A\mathbf{x} = \mathbf{0}$ has three equations and five variables. There must exist free variables. Therefore, $A\mathbf{x} = \mathbf{0}$ has nontrivial solutions, and the columns of A are dependent.

Remark 10 When a matrix A has more columns than rows, the equation $A\mathbf{x} = \mathbf{0}$ must have nontrivial solutions, then we know that the columns of A must be linearly dependent.

Remark 11 If we consider the set of 5 vectors $\left\{ \begin{bmatrix} 1 \\ 5 \\ 3 \end{bmatrix}, \begin{bmatrix} 3 \\ 2 \\ 1 \end{bmatrix}, \begin{bmatrix} 6 \\ 7 \\ 3 \end{bmatrix}, \begin{bmatrix} 1 \\ 1 \\ 1 \end{bmatrix}, \begin{bmatrix} 2 \\ 4 \\ 8 \end{bmatrix} \right\}$ in R^3. These vectors are linearly dependent since the following system

$$x_1 \begin{bmatrix} 1 \\ 5 \\ 3 \end{bmatrix} + x_2 \begin{bmatrix} 3 \\ 2 \\ 1 \end{bmatrix} + x_3 \begin{bmatrix} 6 \\ 7 \\ 3 \end{bmatrix} + x_4 \begin{bmatrix} 1 \\ 1 \\ 1 \end{bmatrix} + x_5 \begin{bmatrix} 2 \\ 4 \\ 8 \end{bmatrix} = \begin{bmatrix} 0 \\ 0 \\ 0 \end{bmatrix}$$

has at least a nontrivial solution. This case can be generalized: If the number of vectors in a set is more than the number of entries in each vector, then the set is linearly dependent. That is, any set $\{\mathbf{a}_1, \ \mathbf{a}_2, \ \ldots, \ \mathbf{a}_n\}$ in R^m with $m < n$ is linearly dependent.

1.7 Exercises

Exercise 12 Decide if each set of vectors is linearly independent. (a) $\left\{ \begin{bmatrix} 1 \\ 2 \end{bmatrix}, \begin{bmatrix} -2 \\ -4 \end{bmatrix} \right\}$,

(b) $\left\{ \begin{bmatrix} 1 \\ 2 \end{bmatrix}, \begin{bmatrix} 3 \\ 8 \end{bmatrix} \right\}$

Exercise 13 Decide if each set of vectors is linearly independent. (a) $\left\{\begin{bmatrix}1\\3\\0\end{bmatrix}, \begin{bmatrix}2\\4\\0\end{bmatrix}\right\}$, (b) $\left\{\begin{bmatrix}-2\\1\\5\end{bmatrix}, \begin{bmatrix}-4\\2\\10\end{bmatrix}\right\}$ (c) $\left\{\begin{bmatrix}1\\3\\0\end{bmatrix}, \begin{bmatrix}2\\4\\7\end{bmatrix}, \begin{bmatrix}0\\0\\0\end{bmatrix}\right\}$

Exercise 14 For each of the following matrix A, determine if the columns of A form a linearly independent set.

$$A = \begin{bmatrix}1 & -3\\-2 & 6\\3 & -9\end{bmatrix}$$

$$A = \begin{bmatrix}1 & 2 & 1\\1 & 1 & -1\\0 & 1 & 2\end{bmatrix}$$

$$A = \begin{bmatrix}-3 & 0 & 12 & -2\\0 & -1 & 4 & 5\end{bmatrix}$$

Exercise 15 Decide whether the set of vectors $\left\{\begin{bmatrix}3\\5\\1\end{bmatrix}, \begin{bmatrix}1\\0\\0\end{bmatrix}, \begin{bmatrix}0\\1\\1\end{bmatrix}, \begin{bmatrix}3\\-1\\7\end{bmatrix}\right\}$ is linearly independent.

Exercise 16 Decide whether the set of vectors $\left\{\begin{bmatrix}4\\2\\3\end{bmatrix}, \begin{bmatrix}2\\-1\\4\end{bmatrix}, \begin{bmatrix}2\\7\\-6\end{bmatrix}\right\}$ is linearly independent.

2 Linear Transformations

2.1 Linear Transformations

Matrix Transformations

Let's first examine a matrix equation

$$A\mathbf{x} = \mathbf{b}$$

where A denotes a matrix with m rows and n columns, $\mathbf{x}$ is a vector in R^n, and $\mathbf{b}$ is a vector in R^m. The equation shows that the vector $\mathbf{x}$ is transformed into the vector $\mathbf{b}$ under the multiplication by the matrix A. The matrix A defines a rule. Solving the matrix equation means finding all vectors $\mathbf{x}$ in R^n that are converted to the vector $\mathbf{b}$ under this rule.

We can adopt T as the name of a **transformation**, which is a rule that maps each vector $\mathbf{x}$ in R^n to a vector $T(\mathbf{x})$ in R^m. We can also call T a mapping or a function. The **domain** of T is R^n and the **codomain** of T is R^m. Each $\mathbf{x}$ is mapped to $T(\mathbf{x})$ under the rule of T, and $T(\mathbf{x})$ is called the **image** of $\mathbf{x}$. The set of all images $T(\mathbf{x})$ is called the **range** of T. In this section, we define $T(\mathbf{x})$ as $A\mathbf{x}$, with A being an m by n matrix.

Example 1 Consider the transformation $T(\mathbf{x}) = A\mathbf{x}$, with $A = \begin{pmatrix} 1 & 0 & 0 \\ 1 & 1 & 1 \end{pmatrix}$. That is, $T : R^3 \to R^2$. For each $\mathbf{x}$ in R^3, we can find its image $T(\mathbf{x})$ in R^2.

$$T(\mathbf{x}) = \begin{pmatrix} 1 & 0 & 0 \\ 1 & 1 & 1 \end{pmatrix} \begin{pmatrix} x_1 \\ x_2 \\ x_3 \end{pmatrix} = \begin{pmatrix} x_1 \\ x_1 + x_2 + x_3 \end{pmatrix}$$

H. Tian, *Linear Algebra*, Synthesis Lectures on Mathematics & Statistics,
https://doi.org/10.1007/978-3-031-84647-2_2

If $\mathbf{x} = \begin{pmatrix} 2 \\ 1 \\ 3 \end{pmatrix}$, the image $T(\mathbf{x}) = \begin{pmatrix} 1 & 0 & 0 \\ 1 & 1 & 1 \end{pmatrix} \begin{pmatrix} 2 \\ 1 \\ 3 \end{pmatrix} = \begin{pmatrix} 2 \\ 6 \end{pmatrix}$.

Example 2 Consider the transformation $T(\mathbf{x}) = A\mathbf{x}$, with $A = \begin{pmatrix} 1 & 0 & 0 \\ 0 & 1 & 0 \end{pmatrix}$. That is, $T : R^3 \to R^2$. For each $\mathbf{x}$ in R^3, we can find its image $T(\mathbf{x})$ in R^2.

$$T(\mathbf{x}) = \begin{pmatrix} 1 & 0 & 0 \\ 0 & 1 & 0 \end{pmatrix} \begin{pmatrix} x_1 \\ x_2 \\ x_3 \end{pmatrix} = \begin{pmatrix} x_1 \\ x_2 \end{pmatrix}$$

Each $\mathbf{x} = \begin{pmatrix} x_1 \\ x_2 \\ x_3 \end{pmatrix}$ in R^3 is mapped to $T(\mathbf{x}) = \begin{pmatrix} x_1 \\ x_2 \end{pmatrix}$ in R^2, and the image vector $T(\mathbf{x})$ is like a shadow projected onto a wall. For example, when $\mathbf{x} = \begin{pmatrix} 2 \\ 1 \\ 3 \end{pmatrix}$, the image vector $T(\mathbf{x}) = \begin{pmatrix} 1 & 0 & 0 \\ 0 & 1 & 0 \end{pmatrix} \begin{pmatrix} 2 \\ 1 \\ 3 \end{pmatrix} = \begin{pmatrix} 2 \\ 1 \end{pmatrix}$.

Example 3 Let $T : R^3 \to R^3$ be a transformation defined by $T(\mathbf{x}) = A\mathbf{x}$, where

$$A = \begin{pmatrix} 1 & 2 & -1 \\ 2 & 3 & 1 \\ 0 & 3 & -8 \end{pmatrix},$$

find all vectors $\mathbf{x}$ with image $T(\mathbf{x})$ being $\mathbf{b} = \begin{pmatrix} 5 \\ 9 \\ 4 \end{pmatrix}$. To answer this question, we should solve $A\mathbf{x} = \mathbf{b}$. That is,

$$\begin{pmatrix} 1 & 2 & -1 \\ 2 & 3 & 1 \\ 0 & 3 & -8 \end{pmatrix} \begin{pmatrix} x_1 \\ x_2 \\ x_3 \end{pmatrix} = \begin{pmatrix} 5 \\ 9 \\ 4 \end{pmatrix}$$

The reduced row echelon form of the augmented matrix

$$\begin{pmatrix} 1 & 2 & -1 & 5 \\ 2 & 3 & 1 & 9 \\ 0 & 3 & -8 & 4 \end{pmatrix} \sim \begin{pmatrix} 1 & 0 & 0 & -2 \\ 0 & 1 & 0 & 4 \\ 0 & 0 & 1 & 1 \end{pmatrix}$$

indicates a unique $\mathbf{x} = \begin{pmatrix} -2 \\ 4 \\ 1 \end{pmatrix}$ whose image under T is the given vector $\mathbf{b}$.

Example 4 Let $T : R^2 \rightarrow R^2$ be a transformation defined by $T(\mathbf{x}) = A\mathbf{x}$, where

$$A = \begin{pmatrix} 1 & 2 \\ 0 & 0 \end{pmatrix},$$

decide if the vector $\mathbf{b} = \begin{pmatrix} 3 \\ 1 \end{pmatrix}$ is in the range of T. To answer this question, we need to determine if $\mathbf{b}$ is the image of some $\mathbf{x}$. The augmented matrix of $A\mathbf{x} = \mathbf{b}$

$$\begin{pmatrix} 1 & 2 & 3 \\ 0 & 0 & 1 \end{pmatrix}$$

indicates there is no solution, which means that the vector $\mathbf{b} = \begin{pmatrix} 3 \\ 1 \end{pmatrix}$ is not image of any $\mathbf{x}$. So $\mathbf{b}$ is not in the range of T.

Linear Transformations

A transformation T is **linear** if the following is true for any scalar c, and for any $\mathbf{u}$ and $\mathbf{v}$ in the domain of T,

$$T(\mathbf{u} + \mathbf{v}) = T(\mathbf{u}) + T(\mathbf{v})$$
$$T(c\mathbf{u}) = cT(\mathbf{u})$$

We can quickly discover that the above two properties imply

$$T(c\mathbf{u} + d\mathbf{v}) = cT(\mathbf{u}) + dT(\mathbf{v})$$

since

$$T(c\mathbf{u} + d\mathbf{v}) = T(c\mathbf{u}) + T(d\mathbf{v}) = cT(\mathbf{u}) + dT(\mathbf{v})$$

When $T(\mathbf{x})$ is defined by $A\mathbf{x}$, the above properties are satisfied,

$$A(\mathbf{u} + \mathbf{v}) = A(\mathbf{u}) + A(\mathbf{v})$$
$$A(c\mathbf{u}) = cA(\mathbf{u})$$

Hence a matrix transformation is linear.

Let's look at the equality $T(c\mathbf{u} + d\mathbf{v}) = cT(\mathbf{u}) + dT(\mathbf{v})$ which is true for any linear transformation T. Its left hand side $T(c\mathbf{u} + d\mathbf{v})$ is the image of the linear combination

$c\mathbf{u} + d\mathbf{v}$, and its right hand side $cT(\mathbf{u}) + dT(\mathbf{v})$ is the linear combination of images of $\mathbf{u}$ and $\mathbf{v}$. By repeatedly applying this property, we arrive at a generalized version for any positive integer k, for any scalars $c_1, c_2, \ldots, c_k$, and for any $\mathbf{u}_1, \mathbf{u}_2, \ldots, \mathbf{u}_k$ in the domain of T,

$$T(c_1\mathbf{u}_1 + c_2\mathbf{u}_2 + \cdots + c_k\mathbf{u}_k) = c_1T(\mathbf{u}_1) + c_2T(\mathbf{u}_2) + \cdots + c_kT(\mathbf{u}_k)$$

Example 5 Consider a linear transformation $T : R^2 \to R^3$. Given $\mathbf{u} = \begin{pmatrix} 1 \\ 0 \end{pmatrix}$, $\mathbf{v} = \begin{pmatrix} 7 \\ -3 \end{pmatrix}$ and their images $T(\mathbf{u}) = \begin{pmatrix} 2 \\ 1 \\ 1 \end{pmatrix}$, $T(\mathbf{v}) = \begin{pmatrix} 1 \\ 4 \\ 5 \end{pmatrix}$, determine the image of $3\mathbf{u} - 2\mathbf{v}$. For this question, we need to find $T(3\mathbf{u} - 2\mathbf{v})$. Let's use the linear property of T and the images of $\mathbf{u}$ and $\mathbf{v}$,

$$T(3\mathbf{u} - 2\mathbf{v}) = 3T(\mathbf{u}) - 2T(\mathbf{v}) = 3\begin{pmatrix} 2 \\ 1 \\ 1 \end{pmatrix} - 2\begin{pmatrix} 1 \\ 4 \\ 5 \end{pmatrix} = \begin{pmatrix} 4 \\ -5 \\ -7 \end{pmatrix}$$

Example 6 Consider a linear transformation $T : R^2 \to R^3$. Given $\mathbf{e}_1 = \begin{pmatrix} 1 \\ 0 \end{pmatrix}$, $\mathbf{e}_2 = \begin{pmatrix} 0 \\ 1 \end{pmatrix}$, and their images $T(\mathbf{e}_1) = \begin{pmatrix} 3 \\ 5 \\ 2 \end{pmatrix}$, $T(\mathbf{e}_2) = \begin{pmatrix} 1 \\ -2 \\ 6 \end{pmatrix}$, decide the image of $\mathbf{u} = \begin{pmatrix} 2 \\ 5 \end{pmatrix}$, and find a formula for the image of an arbitrary $\mathbf{x}$ in R^2. For this question, we first need to find $T(\mathbf{u})$. The given vector $\mathbf{u}$ can be easily represented as a linear combination of $\mathbf{e}_1$ and $\mathbf{e}_2$,

$$\mathbf{u} = \begin{pmatrix} 2 \\ 5 \end{pmatrix} = 2\begin{pmatrix} 1 \\ 0 \end{pmatrix} + 5\begin{pmatrix} 0 \\ 1 \end{pmatrix} = 2\mathbf{e}_1 + 5\mathbf{e}_2$$

The image of $\mathbf{u}$ can be found using the linear property of T and the images of $\mathbf{e}_1$ and $\mathbf{e}_2$,

$$T(\mathbf{u}) = T(2\mathbf{e}_1 + 5\mathbf{e}_2) = 2T(\mathbf{e}_1) + 5T(\mathbf{e}_2) = 2\begin{pmatrix} 3 \\ 5 \\ 2 \end{pmatrix} + 5\begin{pmatrix} 1 \\ -2 \\ 6 \end{pmatrix} = \begin{pmatrix} 11 \\ 0 \\ 34 \end{pmatrix}$$

Similarly, we can find the image of any $\mathbf{x}$ in R^2. The vector $\mathbf{x}$ can be represented as a linear combination of $\mathbf{e}_1$ and $\mathbf{e}_2$,

$$\mathbf{x} = \begin{pmatrix} x_1 \\ x_2 \end{pmatrix} = x_1\begin{pmatrix} 1 \\ 0 \end{pmatrix} + x_2\begin{pmatrix} 0 \\ 1 \end{pmatrix} = x_1\mathbf{e}_1 + x_2\mathbf{e}_2$$

The image of $\mathbf{x}$ is then found using the linear property of T and the images of $\mathbf{e}_1$ and $\mathbf{e}_2$,

$$T(\mathbf{x}) = T(x_1\mathbf{e}_1 + x_2\mathbf{e}_2) = x_1 T(\mathbf{e}_1) + x_2 T(\mathbf{e}_2) = x_1 \begin{pmatrix} 3 \\ 5 \\ 2 \end{pmatrix} + x_2 \begin{pmatrix} 1 \\ -2 \\ 6 \end{pmatrix} = \begin{pmatrix} 3x_1 + x_2 \\ 5x_1 - 2x_2 \\ 2x_1 + 6x_2 \end{pmatrix}$$

which in matrix form is,

$$T(\mathbf{x}) = A\mathbf{x} = \begin{pmatrix} 3 & 1 \\ 5 & -2 \\ 2 & 6 \end{pmatrix} \begin{pmatrix} x_1 \\ x_2 \end{pmatrix}$$

We have found the transformation matrix A, whose first column is $T(\mathbf{e}_1)$ and second column is $T(\mathbf{e}_2)$.

Transformation Matrix

Let us now consider a linear transformation $T : R^n \to R^m$. For the unit vectors $\mathbf{e}_1 = \begin{pmatrix} 1 \\ 0 \\ \vdots \\ 0 \end{pmatrix}$, $\mathbf{e}_2 = \begin{pmatrix} 0 \\ 1 \\ \vdots \\ 0 \end{pmatrix}, \ldots, \mathbf{e}_n = \begin{pmatrix} 0 \\ 0 \\ \vdots \\ 1 \end{pmatrix}$ of R^n, their images in R^m are given respectively as $T(\mathbf{e}_1)$, $T(\mathbf{e}_2), \ldots, T(\mathbf{e}_n)$. Then for any $\mathbf{x}$ in R^n, $\mathbf{x}$ as a linear combination of these unit vectors $\mathbf{e}_j$, $j = 1, 2, \ldots, n$ is as follows,

$$\mathbf{x} = \begin{pmatrix} x_1 \\ x_2 \\ \vdots \\ x_n \end{pmatrix} = x_1 \begin{pmatrix} 1 \\ 0 \\ \vdots \\ 0 \end{pmatrix} + x_2 \begin{pmatrix} 0 \\ 1 \\ \vdots \\ 0 \end{pmatrix} + \cdots + x_n \begin{pmatrix} 0 \\ 0 \\ \vdots \\ 1 \end{pmatrix},$$

i.e.,

$$\mathbf{x} = x_1\mathbf{e}_1 + x_2\mathbf{e}_2 + \cdots + x_n\mathbf{e}_n$$

The image $T(\mathbf{x})$ is then found by using the linear property of T and images of unit vectors,

$$T(\mathbf{x}) = x_1 T(\mathbf{e}_1) + x_2 T(\mathbf{e}_2) + \cdots + x_n T(\mathbf{e}_n)$$

which in matrix form is,

$$T(\mathbf{x}) = [T(\mathbf{e}_1)\ T(\mathbf{e}_2)\ \cdots\ T(\mathbf{e}_n)]\,\mathbf{x}$$

Let us denote the matrix $[T(\mathbf{e}_1)\ T(\mathbf{e}_2)\ \cdots\ T(\mathbf{e}_n)]$ as A. This is the transformation matrix for the linear transformation. The matrix A has n columns with the jth column as $T(\mathbf{e}_j)$,

$j = 1, 2, \ldots, n$. Note that each column $T(\mathbf{e}_j)$ of A is a vector of R^m. The matrix A is m by n. For any $\mathbf{x}$ in R^n, its image can be calculated as $T(\mathbf{x}) = A\mathbf{x}$.

Example 7 Consider a linear transformation $T : R^2 \to R^2$. Given $\mathbf{e}_1 = \begin{pmatrix} 1 \\ 0 \end{pmatrix}$, $\mathbf{e}_2 = \begin{pmatrix} 0 \\ 1 \end{pmatrix}$, and their images $T(\mathbf{e}_1) = \begin{pmatrix} 0 \\ 1 \end{pmatrix}$, $T(\mathbf{e}_2) = \begin{pmatrix} -1 \\ 0 \end{pmatrix}$, determine the transformation matrix.

Using the above analysis, for any $\mathbf{x}$ in R^2,

$$\begin{aligned} T(\mathbf{x}) &= T(x_1\mathbf{e}_1 + x_2\mathbf{e}_2) \\ &= x_1 T(\mathbf{e}_1) + x_2 T(\mathbf{e}_2) \\ &= [T(\mathbf{e}_1)\ T(\mathbf{e}_2)] \begin{pmatrix} x_1 \\ x_2 \end{pmatrix} \\ &= \begin{bmatrix} 0 & -1 \\ 1 & 0 \end{bmatrix} \begin{pmatrix} x_1 \\ x_2 \end{pmatrix} \end{aligned}$$

Therefore, $T(\mathbf{x}) = A\mathbf{x}$, where $A = \begin{bmatrix} 0 & -1 \\ 1 & 0 \end{bmatrix}$. Note that the images of $\mathbf{e}_1$ and $\mathbf{e}_2$ make us realize that this transformation has the effect of rotating a vector in R^2 counterclockwise $90°$ about the origin.

2.1 Exercises

Exercise 8 Let $T : R^3 \to R^2$ be the linear transformation, with $T(\mathbf{x}) = A\mathbf{x}$ given as follows,

$$T(\mathbf{x}) = \begin{pmatrix} 2 & 0 & 1 \\ -3 & 5 & 2 \end{pmatrix} \begin{pmatrix} x_1 \\ x_2 \\ x_3 \end{pmatrix}$$

Decide the image of $\mathbf{x} = \begin{pmatrix} 2 \\ 1 \\ 2 \end{pmatrix}$.

Exercise 9 Let $T : R^2 \to R^4$ be the linear transformation, with $T(\mathbf{x}) = A\mathbf{x}$ given as follows,

$$T(\mathbf{x}) = \begin{pmatrix} 1 & 4 \\ 0 & 1 \\ -2 & 0 \\ 3 & 8 \end{pmatrix} \begin{pmatrix} x_1 \\ x_2 \end{pmatrix}$$

Decide the image of $\mathbf{x} = \begin{pmatrix} 1 \\ 2 \end{pmatrix}$.

Exercise 10 Let $T : R^3 \rightarrow R^2$ be the linear transformation, with $T(\mathbf{x}) = \begin{pmatrix} x_1 + x_3 \\ x_1 + 5x_2 \end{pmatrix}$. Decide the transformation matrix A that $T(\mathbf{x}) = A\mathbf{x}$.

Exercise 11 Let $T : R^2 \rightarrow R^4$ be the linear transformation, with $T(\mathbf{x}) = \begin{pmatrix} x_2 \\ x_1 - x_2 \\ 3x_1 \\ 6x_2 \end{pmatrix}$.

Decide the transformation matrix A that $T(\mathbf{x}) = A\mathbf{x}$.

Exercise 12 Consider a linear transformation $T : R^2 \rightarrow R^3$. Given $\mathbf{u} = \begin{pmatrix} 2 \\ -1 \end{pmatrix}$, $\mathbf{v} = \begin{pmatrix} 3 \\ 3 \end{pmatrix}$ and their images $T(\mathbf{u}) = \begin{pmatrix} 2 \\ 5 \\ 8 \end{pmatrix}$, $T(\mathbf{v}) = \begin{pmatrix} 1 \\ 2 \\ 2 \end{pmatrix}$, determine the image of $2\mathbf{u} - 3\mathbf{v}$.

Exercise 13 Consider a linear transformation $S : R^3 \rightarrow R^2$. Given $\mathbf{u} = \begin{pmatrix} 2 \\ 5 \\ 8 \end{pmatrix}$, $\mathbf{v} = \begin{pmatrix} 1 \\ 2 \\ 2 \end{pmatrix}$ and their images $S(\mathbf{u}) = \begin{pmatrix} 2 \\ -1 \end{pmatrix}$, $S(\mathbf{v}) = \begin{pmatrix} 3 \\ 3 \end{pmatrix}$, determine the image of $\mathbf{u} + 2\mathbf{v}$.

Exercise 14 Let $T : R^3 \rightarrow R^2$ be the linear transformation, with the unit vectors $\mathbf{e}_1 = \begin{pmatrix} 1 \\ 0 \\ 0 \end{pmatrix}$, $\mathbf{e}_2 = \begin{pmatrix} 0 \\ 1 \\ 0 \end{pmatrix}$, $\mathbf{e}_3 = \begin{pmatrix} 0 \\ 0 \\ 1 \end{pmatrix}$ and their images $T(\mathbf{e}_1) = \begin{pmatrix} 1 \\ 3 \end{pmatrix}$, $T(\mathbf{e}_2) = \begin{pmatrix} 4 \\ 2 \end{pmatrix}$, $T(\mathbf{e}_3) = \begin{pmatrix} -1 \\ 5 \end{pmatrix}$. Find the transformation matrix, and the image of $\mathbf{u} = \begin{pmatrix} 7 \\ 8 \\ 9 \end{pmatrix}$.

Exercise 15 Let $T : R^2 \rightarrow R^2$ be the linear transformation defined by rotating a vector counterclockwise 180° about the origin to obtain its image vector. Find the transformation matrix.

Exercise 16 Let $T : R^2 \rightarrow R^2$ be the linear transformation defined by rotating a vector counterclockwise 270° about the origin to obtain its image vector. Find the transformation matrix.

Exercise 17 Let $T : R^2 \to R^2$ be the linear transformation defined by rotating a vector counterclockwise 45° about the origin to obtain its image vector. Find the transformation matrix.

Exercise 18 Let $T : R^2 \to R^2$ be the linear transformation defined by reflecting a vector about the x_1 axis to obtain its image vector. Find the transformation matrix.

Exercise 19 Let $T : R^2 \to R^2$ be the linear transformation defined by reflecting a vector about the x_2 axis to obtain its image vector. Find the transformation matrix.

Exercise 20 Let $T : R^2 \to R^2$ be the linear transformation defined by reflecting a vector about the line $x_2 = x_1$ to obtain its image vector. Find the transformation matrix.

2.2 One-to-One and Onto Transformations

One-to-One Transformations

A transformation (or a mapping) T is called **one-to-one** if $\mathbf{x} \neq \mathbf{y}$ implies $T(\mathbf{x}) \neq T(\mathbf{y})$, that is, distinct input elements correspond to distinct images. For a transformation T, if there exist $\mathbf{x} \neq \mathbf{y}$ with $T(\mathbf{x}) = T(\mathbf{y})$, then the transformation T can not be a one-to-one.

Example 1 Consider the linear transformation $T : R^2 \to R^2$, defined as

$$T(\mathbf{x}) = A\mathbf{x} = \begin{pmatrix} 1 & 2 \\ 0 & 0 \end{pmatrix} \begin{pmatrix} x_1 \\ x_2 \end{pmatrix}$$

Let's determine if T is a one-to-one mapping. We can use the definition of one-to-one mapping to see if $\mathbf{x} \neq \mathbf{y}$ always implies $T(\mathbf{x}) \neq T(\mathbf{y})$. The equation $A\mathbf{x} = \mathbf{b}$ has a free variable. This means that distinct $\mathbf{x}$ vectors can result in the same $\mathbf{b}$ vector. Therefore T is not a one-to-one mapping.

Theorem 2 *Let $T : R^n \to R^m$ be a linear transformation defined by an m by n matrix, $T(\mathbf{x}) = A\mathbf{x}$. The transformation T is one-to-one if and only if $T(\mathbf{x}) = \mathbf{0}$ has only the trivial solution.*

Proof Let's first show that $T(\mathbf{x}) = \mathbf{0}$ has only the trivial solution is a sufficient condition for T being one-to-one. That is, we need to show, by definition, that $\mathbf{u} \neq \mathbf{v}$ implies $T(\mathbf{u}) \neq T(\mathbf{v})$ for the mapping T. This is equivalent to that $T(\mathbf{u}) = T(\mathbf{v})$ implies $\mathbf{u} = \mathbf{v}$. (Recall that the statement "P implies Q" is logically equivalent to "$Not\ Q$ implies $Not\ P$"). With the assumption $T(\mathbf{u}) = T(\mathbf{v})$ and T is linear, then we know

$$T(\mathbf{u}-\mathbf{v}) = T(\mathbf{u}) - T(\mathbf{v}) = \mathbf{0}$$

Since $T(\mathbf{x}) = \mathbf{0}$ has only the trivial solution, that is, $\mathbf{u} - \mathbf{v} = \mathbf{0}$. Therefore $\mathbf{u} = \mathbf{v}$. Next, we show that $T(\mathbf{x}) = \mathbf{0}$ has only the trivial solution is a necessary condition for T being one-to-one. That is, if the linear mapping T is one-to-one, then $T(\mathbf{x}) = \mathbf{0}$ has only the trivial solution. Using the linear property, we have $T(\mathbf{0}) = \mathbf{0}$. Given that T is one-to-one, we know $T(\mathbf{x}) \neq T(\mathbf{0})$ for any $\mathbf{x} \neq \mathbf{0}$, i.e., $T(\mathbf{x}) \neq \mathbf{0}$ for any $\mathbf{x} \neq \mathbf{0}$. Then the only solution for $T(\mathbf{x}) = \mathbf{0}$ is the trivial solution.

Example 3 Consider the linear transformation $T : R^2 \to R^3$, defined as

$$T(\mathbf{x}) = A\mathbf{x} = \begin{pmatrix} 1 & 2 \\ 0 & 1 \\ 0 & 0 \end{pmatrix} \begin{pmatrix} x_1 \\ x_2 \end{pmatrix}$$

Let's determine if T is a one-to-one mapping. We can use the above theorem by checking if $T(\mathbf{x}) = \mathbf{0}$ has only the trivial solution.

$$\begin{pmatrix} 1 & 2 \\ 0 & 1 \\ 0 & 0 \end{pmatrix} \begin{pmatrix} x_1 \\ x_2 \end{pmatrix} = \begin{pmatrix} 0 \\ 0 \\ 0 \end{pmatrix}$$

There is no free variable in the matrix equation $A\mathbf{x} = \mathbf{0}$. Hence there is only a trivial solution $\mathbf{x} = \begin{pmatrix} 0 \\ 0 \end{pmatrix}$. Thus the transformation T is one-to-one.

Theorem 4 *The linear transformation $T : R^n \to R^m$, defined as $T(\mathbf{x}) = A\mathbf{x}$, is one-to-one if and only if the columns of A are linearly independent.* ■

Proof The linear transformation $T(\mathbf{x}) = A\mathbf{x}$ is one-to-one if and only if $T(\mathbf{x}) = \mathbf{0}$ has only the trivial solution. But $T(\mathbf{x}) = \mathbf{0}$ has only the trivial solution if and only if the columns of A are linearly independent. ■

Remark 5 For $T : R^n \to R^m$, with $T(\mathbf{x}) = A\mathbf{x}$. The following statements are equivalent to each other.

(a) The linear transformation $T : R^n \to R^m$, defined as $T(\mathbf{x}) = A\mathbf{x}$, is one-to-one.
(b) The equation $A\mathbf{x} = \mathbf{0}$ has only the trivial solution.
(c) The columns of A are independent.
(d) There is a pivot element in each column of A.

Onto Transformations

Let us consider a linear transformation $T : R^n \rightarrow R^m$, defined as $T(\mathbf{x}) = A\mathbf{x}$, with A being an m by n matrix. Recall that the domain of T is R^n and the codomain of T is R^m. The set of all images $T(\mathbf{x})$ is the range of T. The range of T is a subset of R^m. When the range of T is R^m, the transformation is **onto**. When T is an onto mapping, any vector $\mathbf{b}$ in R^m is an image of some $\mathbf{x}$ in R^n. That is, the equation below has at least a solution $\mathbf{x}$ for any $\mathbf{b}$,

$$A\mathbf{x} = \mathbf{b}$$

This also means any vector $\mathbf{b}$ in R^m can be generated by the columns of A, i.e. R^m is spanned by the columns of A.

Example 6 Consider the linear transformation $T : R^2 \rightarrow R^3$, defined by $T(\mathbf{x}) = A\mathbf{x}$, with $A = \begin{pmatrix} 1 & 0 \\ 0 & 1 \\ 0 & 0 \end{pmatrix}$. Determine if T is onto.

Solution *We consider the matrix equation* $A\mathbf{x} = \mathbf{b}$, *i.e.,*

$$\begin{pmatrix} 1 & 0 \\ 0 & 1 \\ 0 & 0 \end{pmatrix} \begin{pmatrix} x_1 \\ x_2 \end{pmatrix} = \begin{pmatrix} b_1 \\ b_2 \\ b_3 \end{pmatrix}$$

For the transformation T to be onto, the equation $A\mathbf{x} = \mathbf{b}$ should be consistent for any vector $\mathbf{b}$ in R^3. The transformation is not onto if this is not true for some vector $\mathbf{b}$ in R^3. For example, let $\mathbf{b} = \begin{pmatrix} 0 \\ 0 \\ 1 \end{pmatrix}$. The system

$$\begin{pmatrix} 1 & 0 \\ 0 & 1 \\ 0 & 0 \end{pmatrix} \begin{pmatrix} x_1 \\ x_2 \end{pmatrix} = \begin{pmatrix} 0 \\ 0 \\ 1 \end{pmatrix}$$

is obviously inconsistent. This vector $\mathbf{b}$ is not represented by the columns of A, and it is a vector in R^3 which is not an image of any $\mathbf{x}$ in R^2. Therefore, the given transformation is not onto. (As a matter of fact, the columns of A can only generate vectors of the form $\begin{pmatrix} b_1 \\ b_2 \\ 0 \end{pmatrix}$, with third entry being zero.)

Example 7 Consider the linear transformation $T : R^3 \to R^3$, defined by $T(\mathbf{x}) = A\mathbf{x}$, with $A = \begin{pmatrix} 1 & 2 & -1 \\ 2 & 3 & 1 \\ 0 & 3 & -8 \end{pmatrix}$. Determine if T is onto.

Solution *We consider the matrix equation* $A\mathbf{x} = \mathbf{b}$, *i.e.,*

$$\begin{pmatrix} 1 & 2 & -1 \\ 2 & 3 & 1 \\ 0 & 3 & -8 \end{pmatrix}\begin{pmatrix} x_1 \\ x_2 \\ x_3 \end{pmatrix} = \begin{pmatrix} b_1 \\ b_2 \\ b_3 \end{pmatrix}$$

For the transformation T *to be onto, the equation* $A\mathbf{x} = \mathbf{b}$ *should be consistent for any vector* $\mathbf{b}$ *in* R^3. *The matrix* A *can be row reduced to*

$$\begin{pmatrix} 1 & 0 & 0 \\ 0 & 1 & 0 \\ 0 & 0 & 1 \end{pmatrix}$$

which means that there is at least a solution $\mathbf{x}$ *for any* $\mathbf{b}$ *in* R^3. *Therefore, the given transformation is onto.*

Remark 8 For $T : R^n \to R^m$, with $T(\mathbf{x}) = A\mathbf{x}$. The following statements are equivalent to each other.

(a) The linear transformation $T : R^n \to R^m$, defined as $T(\mathbf{x}) = A\mathbf{x}$, is onto.
(b) The equation $A\mathbf{x} = \mathbf{b}$ is consistent for any vector $\mathbf{b}$ in R^m.
(c) R^m is spanned by the columns of A.
(d) There is a pivot element in each row of A.

2.2 Exercises

Exercise 9 Consider $T : R^2 \to R^3$, defined by $T(\mathbf{x}) = A\mathbf{x}$, with $A = \begin{pmatrix} 1 & 0 \\ 2 & 0 \\ 0 & 1 \end{pmatrix}$. (a) Decide if T is one-to-one; (b) Decide if T is onto.

Exercise 10 Consider $T : R^2 \to R^2$, defined by $T(\mathbf{x}) = A\mathbf{x}$, with $A = \begin{pmatrix} -1 & 0 \\ 0 & -1 \end{pmatrix}$. (a) Decide if T is one-to-one; (b) Decide if T is onto.

Exercise 11 Consider $T : R^3 \rightarrow R^3$, defined by $T(\mathbf{x}) = A\mathbf{x}$, with $A = \begin{pmatrix} 1 & 0 & 0 \\ 1 & 2 & 0 \\ 0 & 1 & 1 \end{pmatrix}$.

(a) Decide if T is one-to-one; (b) Decide if T is onto.

Exercise 12 Consider $T : R^4 \rightarrow R^3$, defined by $T(\mathbf{x}) = A\mathbf{x}$, with $A = \begin{pmatrix} 2 & 0 & 0 & 0 \\ 0 & 0 & 1 & 0 \\ 0 & 0 & 0 & 3 \end{pmatrix}$.

(a) Decide if T is one-to-one; (b) Decide if T is onto.

Exercise 13 Consider $T : R^4 \rightarrow R^4$, defined by $T(\mathbf{x}) = A\mathbf{x}$, with $A = \begin{pmatrix} 0 & 0 & 1 & 0 \\ 0 & 0 & 0 & 1 \\ 1 & 0 & 0 & 0 \\ 0 & 1 & 0 & 0 \end{pmatrix}$.

(a) Decide if T is one-to-one; (b) Decide if T is onto.

Exercise 14 Let $T : R^4 \rightarrow R^2$ be the linear transformation, with $T(\mathbf{x}) = \begin{pmatrix} x_1 \\ x_1 + x_2 + x_3 + x_4 \end{pmatrix}$. (a) Decide if T is one-to-one; (b) Decide if T is onto.

Exercise 15 Let $T : R^2 \rightarrow R^4$ be the linear transformation, with $T(\mathbf{x}) = \begin{pmatrix} x_2 \\ x_1 - x_2 \\ 3x_1 \\ 6x_2 \end{pmatrix}$.

(a) Decide if T is one-to-one; (b) Decide if T is onto.

Matrix Algebra 3

3.1 Matrix Operations

Matrix Notations and Definitions

We know that an m by n matrix A is a rectangular array of real numbers. We use $m \times n$ to denote the size of a matrix of m rows and n columns,

$$A = \begin{bmatrix} a_{11} & a_{12} & \cdots & a_{1n} \\ a_{21} & a_{22} & \cdots & a_{2n} \\ \vdots & \vdots & & \vdots \\ a_{m1} & a_{m2} & \cdots & a_{mn} \end{bmatrix}$$

where a_{ij} denotes the entry that lies at the ith row and jth column.

A matrix of $m \times n$ is called a **zero matrix** if each entry is zero.

Example 1 Consider the matrix A,

$$A = \begin{bmatrix} 3 & 1 & 4 \\ -2 & 6 & 5 \end{bmatrix}$$

that has 2 rows and 3 columns. It is a 2×3 matrix. The entries at the first row are $a_{11} = 3$, $a_{12} = 1$, $a_{13} = 4$, and the entries at the second row are $a_{21} = -2$, $a_{22} = 6$, $a_{23} = 5$.

When $m = n$, we have a square matrix $n \times n$. A **diagonal matrix** is an $n \times n$ square matrix with $a_{ij} = 0$ for any $i \neq j$.

H. Tian, *Linear Algebra*, Synthesis Lectures on Mathematics & Statistics,
https://doi.org/10.1007/978-3-031-84647-2_3

Example 2 Here A is a 3×3 diagonal matrix,

$$A = \begin{bmatrix} 5 & 0 & 0 \\ 0 & 2 & 0 \\ 0 & 0 & 8 \end{bmatrix}$$

with its diagonal entries being $a_{11} = 5$, $a_{22} = 2$, $a_{33} = 8$. All the non-diagonal entries $a_{ij} = 0$, for $i \neq j$.

Example 3 Is the following matrix a diagonal matrix?

$$B = \begin{bmatrix} 0 & 0 & 0 \\ 0 & 2 & 0 \\ 0 & 0 & 0 \end{bmatrix}$$

It is a diagonal matrix because all the non-diagonal entries are zeros. That is, all the entries for $i \neq j$, $a_{ij} = 0$.

An **identity matrix** is a diagonal matrix with its diagonal entries being 1. The notation for an $n \times n$ identity matrix is I_n.

Example 4 The identity matrix I_n when $n = 4$.

$$I_4 = \begin{bmatrix} 1 & 0 & 0 & 0 \\ 0 & 1 & 0 & 0 \\ 0 & 0 & 1 & 0 \\ 0 & 0 & 0 & 1 \end{bmatrix}$$

Sums and Scalar Multiples

We can compare two matrices. They are **equal** if and only if they are of the same size and their corresponding entries are equal.

Like how we add two vectors, we can do the sum of two matrices A and B when they are of the same size $m \times n$. We represent them in symbols as,

$$A = \begin{bmatrix} a_{11} & a_{12} & \cdots & a_{1n} \\ a_{21} & a_{22} & \cdots & a_{2n} \\ \vdots & \vdots & & \vdots \\ a_{m1} & a_{m2} & \cdots & a_{mn} \end{bmatrix} \text{ and } B = \begin{bmatrix} b_{11} & b_{12} & \cdots & b_{1n} \\ b_{21} & b_{22} & \cdots & b_{2n} \\ \vdots & \vdots & & \vdots \\ b_{m1} & b_{m2} & \cdots & b_{mn} \end{bmatrix}$$

The **sum** is denoted by $A + B$ whose entries are obtained by adding corresponding entries of A and B. That is,

$$A + B = \begin{bmatrix} a_{11} + b_{11} & a_{12} + b_{12} & \cdots & a_{1n} + b_{1n} \\ a_{21} + b_{21} & a_{22} + b_{22} & \cdots & a_{2n} + b_{2n} \\ \vdots & \vdots & & \vdots \\ a_{m1} + b_{m1} & a_{m2} + b_{m2} & \cdots & a_{mn} + b_{mn} \end{bmatrix}$$

Let c be a scalar and A be a matrix, then **the scalar multiple** cA is the matrix whose entries are c multiple of corresponding entries of A. That is,

$$cA = c \begin{bmatrix} a_{11} & a_{12} & \cdots & a_{1n} \\ a_{21} & a_{22} & \cdots & a_{2n} \\ \vdots & \vdots & & \vdots \\ a_{m1} & a_{m2} & \cdots & a_{mn} \end{bmatrix} = \begin{bmatrix} ca_{11} & ca_{12} & \cdots & ca_{1n} \\ ca_{21} & ca_{22} & \cdots & ca_{2n} \\ \vdots & \vdots & & \vdots \\ ca_{m1} & ca_{m2} & \cdots & ca_{mn} \end{bmatrix}$$

The difference $A - B$ is the same as $A + (-1)B$.

Example 5 Given

$$A = \begin{bmatrix} 2 & 0 & -3 \\ 1 & 6 & 4 \end{bmatrix} \text{ and } B = \begin{bmatrix} 2 & 3 & 0 \\ 1 & -1 & 2 \end{bmatrix},$$

find $2A$ and $2A + B$. The scalar multiple $2A$ is calculated as,

$$2A = 2 \begin{bmatrix} 2 & 0 & -3 \\ 1 & 6 & 4 \end{bmatrix} = \begin{bmatrix} 4 & 0 & -6 \\ 2 & 12 & 8 \end{bmatrix}$$

and $2A + B$ is calculated as,

$$2A + B = \begin{bmatrix} 4 & 0 & -6 \\ 2 & 12 & 8 \end{bmatrix} + \begin{bmatrix} 2 & 3 & 0 \\ 1 & -1 & 2 \end{bmatrix} = \begin{bmatrix} 6 & 3 & -6 \\ 3 & 11 & 10 \end{bmatrix}$$

The following properties can be shown to be true for the $m \times n$ matrices $A, B, C, 0$, and the scalars c and d.

1. $A + B = B + A$
2. $(A + B) + C = A + (B + C)$
3. $A + 0 = 0 + A = A$
4. There is a matrix $-A$ such that $A + (-A) = 0$
5. $c(A + B) = cA + cB$
6. $(c + d)A = cA + dA$
7. $c(dA) = (cd)A$
8. $1A = A$

Matrix Multiplication

Recall that an $m \times n$ matrix B can define a linear transformation that maps a vector $\mathbf{x}$ in R^n to the $B\mathbf{x}$ in R^m, i.e.,

$$\mathbf{x} \rightarrow B\mathbf{x}$$

Suppose we consider another linear transformation defined by a $k \times m$ matrix A, that maps a vector $\mathbf{u}$ in R^m to the $A\mathbf{u}$ in R^k, i.e.,

$$\mathbf{u} \rightarrow A\mathbf{u}$$

Then we can apply the first transformation followed by the second transformation,

$$\mathbf{x} \rightarrow B\mathbf{x} \rightarrow A\left(B\mathbf{x}\right)$$

So $A\left(B\mathbf{x}\right)$ can be viewed as the image of $\mathbf{x}$, by a new transformation $R^n \rightarrow R^k$,

$$\mathbf{x} \rightarrow A\left(B\mathbf{x}\right)$$

Regarding this new transformation, we know how to find the image of any $\mathbf{x}$ in R^n. This is specifically how: take $\mathbf{x}$ in R^n, we first find $B\mathbf{x}$ in R^m and next $A\left(B\mathbf{x}\right)$ in R^k. The new transformation is the composition of two linear transformations defined by matrix A and B. Let AB denotes the transformation matrix of the composite transformation, then the matrix AB should satisfy the property

$$AB\left(\mathbf{x}\right) = A\left(B\mathbf{x}\right)$$

In the following, let's discover how to properly define the product matrix AB so that it corresponds to the composite transformation.

Recall that the transformation matrix can be determined by the images of the unit vectors in the domain R^n. Let us denote the unit vectors in R^n by

$$\mathbf{e_1} = \begin{bmatrix} 1 \\ 0 \\ \vdots \\ 0 \end{bmatrix}, \mathbf{e_2} = \begin{bmatrix} 0 \\ 1 \\ \vdots \\ 0 \end{bmatrix}, \ldots, \mathbf{e_n} = \begin{bmatrix} 0 \\ 0 \\ \vdots \\ 1 \end{bmatrix},$$

then the image of $\mathbf{e_1}$ using the calculation of $A\left(B\mathbf{e_1}\right)$,

$$B\mathbf{e_1} = \begin{bmatrix} b_{11} & b_{12} & \cdots & b_{1n} \\ b_{21} & b_{22} & \cdots & b_{2n} \\ \vdots & \vdots & & \vdots \\ b_{m1} & b_{m2} & \cdots & b_{mn} \end{bmatrix} \begin{bmatrix} 1 \\ 0 \\ \vdots \\ 0 \end{bmatrix} = \begin{bmatrix} b_{11} \\ b_{21} \\ \vdots \\ b_{m1} \end{bmatrix} = \mathbf{b}_1$$

where $\mathbf{b}_1$ denotes the matrix B's first column. Thus,

$$A\left(B\mathbf{e_1}\right) = A\mathbf{b}_1$$

Similarly, the image of $\mathbf{e_j}$ using the calculation of $A\left(B\mathbf{e_j}\right)$,

$$B\mathbf{e_j} = \begin{bmatrix} b_{11} & b_{12} & \cdots & b_{1n} \\ b_{21} & b_{22} & \cdots & b_{2n} \\ \vdots & \vdots & & \vdots \\ b_{m1} & b_{m2} & \cdots & b_{mn} \end{bmatrix} \mathbf{e_j} = \begin{bmatrix} b_{1j} \\ b_{2j} \\ \vdots \\ b_{mj} \end{bmatrix} = \mathbf{b_j}$$

where $\mathbf{b_j}$ denotes the matrix B's jth column. Therefore,

$$A\left(B\mathbf{e_j}\right) = A\mathbf{b_j}$$

for $j = 1, 2, \ldots, n$. The transformation matrix AB that defines the composite transformation $A(B\mathbf{x}) : R^n \to R^k$ is,

$$AB = [A\mathbf{b}_1 \; A\mathbf{b}_2 \; \cdots \; A\mathbf{b}_n]$$

which is a $k \times n$ matrix. That is, $A(B\mathbf{x}) = AB(\mathbf{x})$. We define the multiplication of two matrices, a $k \times m$ matrix A and an $m \times n$ matrix B, as follows,

$$\begin{aligned} AB &= A\,[\mathbf{b}_1 \; \mathbf{b}_2 \; \cdots \; \mathbf{b}_n] \\ &= [A\mathbf{b}_1 \; A\mathbf{b}_2 \; \cdots \; A\mathbf{b}_n] \end{aligned}$$

That is, the resulting matrix AB is a $k \times n$ matrix, with its jth column being $A\mathbf{b_j}$, $j = 1, 2, \ldots, n$.

Example 6 Let $A = \begin{bmatrix} 1 & 0 \\ 0 & 1 \\ 1 & 0 \end{bmatrix}$ and $B = \begin{bmatrix} 2 & 1 & 0 & 1 \\ 0 & 1 & 1 & 0 \end{bmatrix}$, decide AB.

Solution *The matrix B has 4 columns* $\mathbf{b}_1 = \begin{bmatrix} 2 \\ 0 \end{bmatrix}$, $\mathbf{b}_2 = \begin{bmatrix} 1 \\ 1 \end{bmatrix}$, $\mathbf{b}_3 = \begin{bmatrix} 0 \\ 1 \end{bmatrix}$, $\mathbf{b}_4 = \begin{bmatrix} 1 \\ 0 \end{bmatrix}$. *We compute the 4 columns of AB as follows,*

$$A\mathbf{b}_1 = \begin{bmatrix} 1 & 0 \\ 0 & 1 \\ 1 & 0 \end{bmatrix} \begin{bmatrix} 2 \\ 0 \end{bmatrix} = \begin{bmatrix} 2 \\ 0 \\ 2 \end{bmatrix}$$

$$A\mathbf{b}_2 = \begin{bmatrix} 1 & 0 \\ 0 & 1 \\ 1 & 0 \end{bmatrix} \begin{bmatrix} 1 \\ 1 \end{bmatrix} = \begin{bmatrix} 1 \\ 1 \\ 1 \end{bmatrix}$$

$$A\mathbf{b}_3 = \begin{bmatrix} 1 & 0 \\ 0 & 1 \\ 1 & 0 \end{bmatrix} \begin{bmatrix} 0 \\ 1 \end{bmatrix} = \begin{bmatrix} 0 \\ 1 \\ 0 \end{bmatrix}$$

$$\mathbf{Ab}_4 = \begin{bmatrix} 1 & 0 \\ 0 & 1 \\ 1 & 0 \end{bmatrix} \begin{bmatrix} 1 \\ 0 \end{bmatrix} = \begin{bmatrix} 1 \\ 0 \\ 1 \end{bmatrix}$$

Therefore,

$$AB = [\mathbf{Ab}_1 \ \ \mathbf{Ab}_2 \ \ \mathbf{Ab}_3 \ \ \mathbf{Ab}_4]$$
$$= \begin{bmatrix} 2 & 1 & 0 & 1 \\ 0 & 1 & 1 & 0 \\ 2 & 1 & 0 & 1 \end{bmatrix}$$

Remark 7 Each column of the matrix AB is a linear combination of the columns of A, with weights being entries of the corresponding column of B.

Using the above approach, we can compute all columns of AB. This leads to an alternative formula for the entries of AB. Let's start with the jth column of AB, i.e.,

$$\mathbf{Ab}_j = \begin{bmatrix} a_{11} & a_{12} & \cdots & a_{1m} \\ a_{21} & a_{22} & \cdots & a_{2m} \\ \vdots & \vdots & & \vdots \\ a_{k1} & a_{k2} & \cdots & a_{km} \end{bmatrix} \begin{bmatrix} b_{1j} \\ b_{2j} \\ \vdots \\ b_{mj} \end{bmatrix}$$

which can be written as a linear combination of the columns of A, i.e.,

$$\mathbf{Ab}_j = b_{1j} \begin{bmatrix} a_{11} \\ a_{21} \\ \vdots \\ a_{k1} \end{bmatrix} + b_{2j} \begin{bmatrix} a_{12} \\ a_{22} \\ \vdots \\ a_{k2} \end{bmatrix} + \cdots + b_{mj} \begin{bmatrix} a_{1m} \\ a_{2m} \\ \vdots \\ a_{km} \end{bmatrix}$$

After combining these vectors, we obtain

$$\mathbf{Ab}_j = \begin{bmatrix} a_{11}b_{1j} + a_{12}b_{2j} + \cdots + a_{1m}b_{mj} \\ a_{21}b_{1j} + a_{22}b_{2j} + \cdots + a_{2m}b_{mj} \\ \vdots \\ a_{k1}b_{1j} + a_{k2}b_{2j} + \cdots + a_{km}b_{mj} \end{bmatrix}$$

The entry of AB at the ith row and jth column is just the ith entry of the $\mathbf{Ab}_j$, and it is calculated as,

$$(AB)_{ij} = a_{i1}b_{1j} + a_{i2}b_{2j} + \cdots + a_{im}b_{mj}$$

which involves the ith row of A and jth column of B.

Example 8 Let $A = \begin{bmatrix} 1 & 2 \\ 0 & 1 \\ 1 & 0 \end{bmatrix}$ and $B = \begin{bmatrix} 2 & 1 & 0 & 1 \\ 0 & 1 & 5 & 0 \end{bmatrix}$, decide AB.

Solution *The size of matrix A is* 3×2 *and the size of the matrix B is* 2×4. *Thus the size of* AB *is* 3×4. *Let's denote* AB *by* C, *then*

$$C = \begin{bmatrix} c_{11} & c_{12} & c_{13} & c_{14} \\ c_{21} & c_{22} & c_{23} & c_{24} \\ c_{31} & c_{32} & c_{33} & c_{34} \end{bmatrix}$$

The entry $c_{ij} = (AB)_{ij}$ *is determined using the* i*th row of* A *and* j*th column of* B. *For example, the value of* c_{13} *depends only on the first row of* A *and the third column of* B,

$$AB = \begin{bmatrix} 1 & 2 \\ 0 & 1 \\ 1 & 0 \end{bmatrix} \begin{bmatrix} 2 & 1 & 0 & 1 \\ 0 & 1 & 5 & 0 \end{bmatrix} = \begin{bmatrix} c_{11} & c_{12} & c_{13} & c_{14} \\ c_{21} & c_{22} & c_{23} & c_{24} \\ c_{31} & c_{32} & c_{33} & c_{34} \end{bmatrix}$$

$$c_{13} = a_{11}b_{13} + a_{12}b_{23} = 1\,(0) + 2\,(5) = 10$$

Similarly, we can find all the entries c_{ij},

$$AB = \begin{bmatrix} 1 & 2 \\ 0 & 1 \\ 1 & 0 \end{bmatrix} \begin{bmatrix} 2 & 1 & 0 & 1 \\ 0 & 1 & 5 & 0 \end{bmatrix}$$

$$= \begin{bmatrix} 2 & 3 & 10 & 1 \\ 0 & 1 & 5 & 0 \\ 2 & 1 & 0 & 1 \end{bmatrix}$$

Remark 9 Using the matrices A and B given in the above example, can we compute BA? The answer is no. The matrix A is 3×2 and the matrix B is 2×4. We can do AB since the number of columns of A is the same as the number of rows of B. We can't do BA since B has 4 columns and A has 3 rows. When the column number of the first matrix is not equal to the row number of the second matrix, the multiplication is undefined. That is, there is no commutative property for matrix multiplication. For some matrices A and B when their dimensions match, you can possibly compute both AB and BA. However, AB and BA are not necessarily equal.

The following **properties** can be proved for matrices A, B, C with the assumption that their dimensions allow multiplication.

1. $(AB)\,C = A\,(BC)$
2. $A\,(B + C) = AB + AC$
3. $(A + B)\,C = AC + BC$
4. $c\,(AB) = (cA)\,B = A\,(cB)$, for a scalar c
5. $I_m A = A_{m\times n} = AI_n$

Remark 10 When $AB = \mathbf{0}$, it is not necessarily true that $A = \mathbf{0}$ or $B = \mathbf{0}$. Here is a simple counter example with $A = \begin{bmatrix} 1 & 0 \\ 0 & 0 \end{bmatrix} \neq \mathbf{0}$ and $B = \begin{bmatrix} 0 & 0 \\ 0 & 1 \end{bmatrix} \neq \mathbf{0}$, but $AB = \begin{bmatrix} 0 & 0 \\ 0 & 0 \end{bmatrix} = \mathbf{0}$.

Remark 11 When $AB = AC$, we should not simply cancel A to conclude that $B = C$. This is because $AB = AC$ implies $AB - AC = \mathbf{0}$, which is equivalent to $A\,(B - C) = \mathbf{0}$. When the product of A and $B - C$ is $\mathbf{0}$, it is not necessarily true that $A = 0$, or $B - C = 0$. Hence, it is not necessarily true that $B = C$.

Further Interpretation of Matrix Multiplication

Let us consider again the composite function $A\,(B\mathbf{x})$, with the assumption that A is $k \times m$, and B is $m \times n$. It maps a vector $\mathbf{x}$ in R^n to a vector in R^k, i.e.,

$$\mathbf{x} \to A\,(B\mathbf{x})$$

Let $\mathbf{x} = \begin{bmatrix} x_1 \\ x_2 \\ \vdots \\ x_n \end{bmatrix}$, then

$$\begin{aligned} B\mathbf{x} &= [\mathbf{b}_1\ \mathbf{b}_2\ \cdots\ \mathbf{b}_n] \begin{bmatrix} x_1 \\ x_2 \\ \vdots \\ x_n \end{bmatrix} \\ &= x_1\mathbf{b}_1 + x_2\mathbf{b}_2 + \cdots + x_n\mathbf{b}_n \end{aligned}$$

which is a linear combination of the columns of B. Using the property of linear transformation,

$$\begin{aligned} A\,(B\mathbf{x}) &= A\,(x_1\mathbf{b}_1 + x_2\mathbf{b}_2 + \cdots + x_n\mathbf{b}_n) \\ &= x_1 A\mathbf{b}_1 + x_2 A\mathbf{b}_2 + \cdots + x_n A\mathbf{b}_n \end{aligned}$$

which is a linear combination of the column vectors $A\mathbf{b}_1, A\mathbf{b}_2, \ldots, A\mathbf{b}_n$. Equivalently,

$$A(B\mathbf{x}) = [A\mathbf{b}_1 \ A\mathbf{b}_2 \ \cdots \ A\mathbf{b}_n] \begin{bmatrix} x_1 \\ x_2 \\ \vdots \\ x_n \end{bmatrix}$$

So the transformation matrix for the mapping $A(B\mathbf{x})$ is $[A\mathbf{b}_1 \ A\mathbf{b}_2 \ \cdots \ A\mathbf{b}_n]$, denoted by AB. Thus, we have verified that the matrix AB defines the same mapping due to the fact that $A(B\mathbf{x}) = AB(\mathbf{x})$

The Transpose

The transpose of a matrix A is denoted by A^T, the row entries of A become the corresponding column entries of A^T. When the size of A is $m \times n$, the size of A^T is $n \times m$.

Example 12 Let $A = \begin{bmatrix} 5 & -3 & 1 \\ 2 & 3 & 4 \end{bmatrix}$, then

$$A^T = \begin{bmatrix} 5 & 2 \\ -3 & 3 \\ 1 & 4 \end{bmatrix}$$

The following basic properties of the transpose operation can be proved:

1. $(A^T)^T = A$
2. $(cA)^T = cA^T$, for any scalar c
3. $(A + B)^T = A^T + B^T$
4. $(AB)^T = B^T A^T$

3.1 Exercises

Exercise 13 Let $A = \begin{bmatrix} 6 & 0 & 2 \\ 0 & -3 & 1 \end{bmatrix}$ and $B = \begin{bmatrix} 1 & 0 & 2 \\ 2 & 1 & 1 \end{bmatrix}$. Find (1) $3B$ (2) $A - 2B$

Exercise 14 Let $A = \begin{bmatrix} 2 & 1 \\ 3 & 4 \end{bmatrix}$, $B = \begin{bmatrix} 1 & 0 & 2 \\ 2 & 1 & 1 \end{bmatrix}$, and $C = \begin{bmatrix} 3 & 0 \\ 0 & 1 \\ 1 & 5 \end{bmatrix}$. Compute each expression or explain why it is not defined.

(1) AB (2) BA (3) BC (4) CB (5) AC (6) CA

Exercise 15 Let A be an $m \times n$ matrix. What must be true for the product AA to be defined?

Exercise 16 Find $A^3 = AAA$, given $A = \begin{bmatrix} -2 & 1 \\ 1 & 2 \end{bmatrix}$.

Exercise 17 Let $A = \begin{bmatrix} 1 & 0 \\ 0 & 3 \\ 2 & -1 \end{bmatrix}$. Compute each expression or explain why it is not defined.

$$(1)\ A^T \quad (2)\ \left(A^T\right)^T \quad (3)\ A^T A \quad (4)\ AA^T$$

Exercise 18 Let $A = \begin{pmatrix} 2 & 0 & 0 & 1 \\ 3 & 0 & 1 & 0 \end{pmatrix}$. Compute each expression or explain why it is not defined.

$$(1)\ I_2 A \quad (2)\ I_4 A \quad (3)\ AI_2 \quad (4)\ AI_4$$

Exercise 19 Let $A = \begin{bmatrix} 5 & 0 \\ 0 & 2 \\ 2 & -1 \end{bmatrix}$ and $C = \begin{bmatrix} 1 & 0 & 3 \\ 6 & 1 & 0 \end{bmatrix}$. Compute each expression or explain why it is not defined.

$$(1)\ A^T - C \quad (2)\ C^T + 2A \quad (3)\ A^T C \quad (4)\ C^T C$$

3.2 Matrix Inverse

Invertible Matrices

An $n \times n$ matrix A is invertible if there is an $n \times n$ matrix B that satisfies

$$AB = I_n \text{ and } BA = I_n$$

The matrix B is called the **inverse** of A and it is usually denoted by A^{-1}.

Example 1 Let $A = \begin{bmatrix} 2 & 5 \\ -3 & -7 \end{bmatrix}$ and $B = \begin{bmatrix} -7 & -5 \\ 3 & 2 \end{bmatrix}$. Then

$$AB = \begin{bmatrix} 2 & 5 \\ -3 & -7 \end{bmatrix}\begin{bmatrix} -7 & -5 \\ 3 & 2 \end{bmatrix} = \begin{bmatrix} 1 & 0 \\ 0 & 1 \end{bmatrix}$$

and

$$BA = \begin{bmatrix} -7 & -5 \\ 3 & 2 \end{bmatrix}\begin{bmatrix} 2 & 5 \\ -3 & -7 \end{bmatrix} = \begin{bmatrix} 1 & 0 \\ 0 & 1 \end{bmatrix}.$$

So B is the inverse matrix of A. The matrix A is also the inverse matrix of B.

Example 2 Consider $A\mathbf{x} = \mathbf{b}$, with $A = \begin{bmatrix} 2 & 5 \\ -3 & -7 \end{bmatrix}$ and $\mathbf{b} = \begin{bmatrix} 1 \\ 6 \end{bmatrix}$. Find $\mathbf{x}$.

Solution *From previous example, we know that the matrix A is invertible and $A^{-1} = B = \begin{bmatrix} -7 & -5 \\ 3 & 2 \end{bmatrix}$. Then we can solve $A\mathbf{x} = \mathbf{b}$ by multiplying both sides of the equation by A^{-1}, i.e.,*

$$A^{-1}(A\mathbf{x}) = A^{-1}\mathbf{b}$$

which, by associative property, is,

$$\left(A^{-1}A\right)\mathbf{x} = A^{-1}\mathbf{b}$$

which yields,

$$I_2\mathbf{x} = A^{-1}\mathbf{b}$$

Therefore

$$\mathbf{x} = A^{-1}\mathbf{b} = \begin{bmatrix} -7 & -5 \\ 3 & 2 \end{bmatrix}\begin{bmatrix} 1 \\ 6 \end{bmatrix} = \begin{bmatrix} -37 \\ 15 \end{bmatrix}$$

For a general 2×2 matrix A, there is a convenient formula for the inverse when it exists. The following Theorem allows us to quickly determine if a 2×2 matrix is invertible, and when it is, to find its inverse.

Theorem 3 *Let $A = \begin{bmatrix} a & b \\ c & d \end{bmatrix}$. The quantity $ad - bc$ is called the determinant of A, denoted by $\det(A)$. When $\det(A) \neq 0$, the matrix A is invertible and its inverse*

$$A^{-1} = \frac{1}{\det(A)}\begin{bmatrix} d & -b \\ -c & a \end{bmatrix}.$$

Proof It can be verified using the definition of inverse matrix. ∎

Example 4 Decide if $A = \begin{bmatrix} 1 & 2 \\ 3 & 2 \end{bmatrix}$ is invertible. If it is, find its inverse.

Solution *The determinant of A is*

$$\det(A) = 1(2) - 2(3) = -4 \neq 0$$

So A is invertible and its inverse, by the formula, is

$$
\begin{aligned}
A^{-1} &= \frac{1}{\det(A)}\begin{bmatrix} d & -b \\ -c & a \end{bmatrix} \\
&= -\frac{1}{4}\begin{bmatrix} 2 & -2 \\ -3 & 1 \end{bmatrix} \\
&= \begin{bmatrix} -\frac{1}{2} & \frac{1}{2} \\ \frac{3}{4} & -\frac{1}{4} \end{bmatrix}
\end{aligned}
$$

Theorem 5 *If A is an $n \times n$ invertible matrix, the equation $A\mathbf{x} = \mathbf{b}$ has a unique solution* $\mathbf{x} = A^{-1}\mathbf{b}$.

Proof Let's solve $A\mathbf{x} = \mathbf{b}$ by multiplying both sides of the equation by A^{-1}, i.e.,

$$A^{-1}(A\mathbf{x}) = A^{-1}\mathbf{b}$$

which, by associative property, is,

$$\left(A^{-1}A\right)\mathbf{x} = A^{-1}\mathbf{b}$$

which yields,

$$I_n\mathbf{x} = A^{-1}\mathbf{b}$$

Therefore

$$\mathbf{x} = A^{-1}\mathbf{b}$$

To check that $A^{-1}\mathbf{b}$ is truly a solution, we substitute it into the left hand of the equation $A\mathbf{x} = \mathbf{b}$,

$$A\mathbf{x} = A\left(A^{-1}\mathbf{b}\right) = \left(AA^{-1}\right)\mathbf{b} = I_n\mathbf{b} = \mathbf{b}$$

which is the same as the right hand side. This proves that the system $A\mathbf{x} = \mathbf{b}$ has a unique solution $\mathbf{x} = A^{-1}\mathbf{b}$ for an invertible matrix A. ■

The invertible matrices are also called **nonsingular** matrices. The **singular** matrices are those that are not invertible.

Let both A and B be $n \times n$ matrices. The following properties can be proved.

Property (1). If $AB = I_n$, then $B = A^{-1}$ and $A = B^{-1}$.

Property (2). If A is invertible, then $\left(A^{-1}\right)^{-1} = A$.

Property (3). If A and B are both invertible, then their product AB is invertible, and $(AB)^{-1} = B^{-1}A^{-1}$.

Property (4). If A is invertible, then A^T is invertible, and $\left(A^T\right)^{-1} = \left(A^{-1}\right)^T$.

Here, let's only prove the property (3) by checking that the matrix $B^{-1}A^{-1}$ is indeed the inverse of the matrix AB. That is, we want to verify that $(AB)\left(B^{-1}A^{-1}\right) = I_n$ and $\left(B^{-1}A^{-1}\right)(AB) = I_n$. We can easily verify each identity using associative property of matrix multiplication:

$$(AB)\left(B^{-1}A^{-1}\right) = A\left(BB^{-1}\right)A^{-1} = A\left(I_n\right)A^{-1} = AA^{-1} = I_n$$

and

$$\left(B^{-1}A^{-1}\right)(AB) = B^{-1}\left(A^{-1}A\right)B = B^{-1}\left(I_n\right)B = B^{-1}B = I_n.$$

Inverse of an Elementary Matrix

An elementary matrix is a square matrix obtained from a single elementary row operation on an identity matrix. Recall that we have three different elementary row operations on any matrix.

Example 6 The following 3×3 matrices E_1, E_2 and E_3 are elementary matrices when we compare each with the 3×3 identity matrix $I = \begin{bmatrix} 1 & 0 & 0 \\ 0 & 1 & 0 \\ 0 & 0 & 1 \end{bmatrix}$,

$$E_1 = \begin{bmatrix} 0 & 0 & 1 \\ 0 & 1 & 0 \\ 1 & 0 & 0 \end{bmatrix}$$

results from the swapping the first row and the third row of the identity matrix;

$$E_2 = \begin{bmatrix} 1 & 0 & 0 \\ 0 & 3 & 0 \\ 0 & 0 & 1 \end{bmatrix}$$

results from multiplying the second row of the identity matrix by 3; and

$$E_3 = \begin{bmatrix} 1 & 0 & 2 \\ 0 & 1 & 0 \\ 0 & 0 & 1 \end{bmatrix}$$

results from replacing the first row of the identity matrix with the sum of its first row and twice its third row.

Example 7 Using the 3×3 elementary matrices E_1, E_2 and E_3 given in the previous example, calculate the products E_1A, E_2A, and E_3A for a general 3×3 matrix A.

Solution *First, we denote a general* 3×3 *matrix* A *by*

$$A = \begin{bmatrix} a_{11} & a_{12} & a_{13} \\ a_{21} & a_{22} & a_{23} \\ a_{31} & a_{32} & a_{33} \end{bmatrix}$$

Next, we calculate each product,

$$E_1A = \begin{bmatrix} 0 & 0 & 1 \\ 0 & 1 & 0 \\ 1 & 0 & 0 \end{bmatrix} \begin{bmatrix} a_{11} & a_{12} & a_{13} \\ a_{21} & a_{22} & a_{23} \\ a_{31} & a_{32} & a_{33} \end{bmatrix}$$

$$= \begin{bmatrix} a_{31} & a_{32} & a_{33} \\ a_{21} & a_{22} & a_{23} \\ a_{11} & a_{12} & a_{13} \end{bmatrix}$$

which is the same matrix when we swap the first row and the third row of A;

$$E_2A = \begin{bmatrix} 1 & 0 & 0 \\ 0 & 3 & 0 \\ 0 & 0 & 1 \end{bmatrix} \begin{bmatrix} a_{11} & a_{12} & a_{13} \\ a_{21} & a_{22} & a_{23} \\ a_{31} & a_{32} & a_{33} \end{bmatrix}$$

$$= \begin{bmatrix} a_{11} & a_{12} & a_{13} \\ 3a_{21} & 3a_{22} & 3a_{23} \\ a_{31} & a_{32} & a_{33} \end{bmatrix}$$

which is the same matrix when we multiply the second row of A *by* 3;

$$E_3A = \begin{bmatrix} 1 & 0 & 2 \\ 0 & 1 & 0 \\ 0 & 0 & 1 \end{bmatrix} \begin{bmatrix} a_{11} & a_{12} & a_{13} \\ a_{21} & a_{22} & a_{23} \\ a_{31} & a_{32} & a_{33} \end{bmatrix}$$

$$= \begin{bmatrix} a_{11} + 2a_{31} & a_{12} + 2a_{32} & a_{13} + 2a_{33} \\ a_{21} & a_{22} & a_{23} \\ a_{31} & a_{32} & a_{33} \end{bmatrix}$$

which is the same matrix when we replace the first row of A *with the sum of its first row and twice its third row.*

Remark 8 When an elementary row operation is applied to a general $m \times n$ matrix A, the resulting matrix can be expressed as the product of an elementary matrix and A. The elementary matrix is the one from the same row operation on the identity matrix.

The inverse of an elementary matrix is easy to find.

Example 9 For each elementary matrix (E_1, E_2, and E_3) in the previous example, verify its inverse matrix as given below

$$E_1^{-1} = \begin{bmatrix} 0 & 0 & 1 \\ 0 & 1 & 0 \\ 1 & 0 & 0 \end{bmatrix},$$

$$E_2^{-1} = \begin{bmatrix} 1 & 0 & 0 \\ 0 & \frac{1}{3} & 0 \\ 0 & 0 & 1 \end{bmatrix}$$

and

$$E_3^{-1} = \begin{bmatrix} 1 & 0 & -2 \\ 0 & 1 & 0 \\ 0 & 0 & 1 \end{bmatrix}$$

By Property (1), we only need to show $AB = I_n$ in order to prove B is the inverse of A, or A is the inverse of B. For the matrix E_1, its inverse is itself since

$$\begin{bmatrix} 0 & 0 & 1 \\ 0 & 1 & 0 \\ 1 & 0 & 0 \end{bmatrix} \begin{bmatrix} 0 & 0 & 1 \\ 0 & 1 & 0 \\ 1 & 0 & 0 \end{bmatrix} = \begin{bmatrix} 1 & 0 & 0 \\ 0 & 1 & 0 \\ 0 & 0 & 1 \end{bmatrix}$$

Similarly, the inverse of E_2 can be verified as,

$$\begin{bmatrix} 1 & 0 & 0 \\ 0 & 3 & 0 \\ 0 & 0 & 1 \end{bmatrix} \begin{bmatrix} 1 & 0 & 0 \\ 0 & \frac{1}{3} & 0 \\ 0 & 0 & 1 \end{bmatrix} = \begin{bmatrix} 1 & 0 & 0 \\ 0 & 1 & 0 \\ 0 & 0 & 1 \end{bmatrix}$$

and the inverse of E_3 can be verified as,

$$\begin{bmatrix} 1 & 0 & 2 \\ 0 & 1 & 0 \\ 0 & 0 & 1 \end{bmatrix} \begin{bmatrix} 1 & 0 & -2 \\ 0 & 1 & 0 \\ 0 & 0 & 1 \end{bmatrix} = \begin{bmatrix} 1 & 0 & 0 \\ 0 & 1 & 0 \\ 0 & 0 & 1 \end{bmatrix}$$

Remark 10 A general case is still true. It can be shown that any elementary matrix is invertible and the inverse of an elementary matrix is also an elementary matrix.

An Approach to Finding Inverse

Example 11 If possible, reduce the matrix $A = \begin{bmatrix} 3 & 0 & 15 \\ 0 & 8 & 1 \\ 0 & 1 & 0 \end{bmatrix}$ to an identity matrix, and express A as the product of elementary matrices.

Solution *Let's simply apply an elementary row operation at each step. More specifically, Step (1), scaling of the first row, replacement of the second row,*

$$A = \begin{bmatrix} 3 & 0 & 15 \\ 0 & 8 & 1 \\ 0 & 1 & 0 \end{bmatrix} \rightarrow \begin{bmatrix} 1 & 0 & 5 \\ 0 & 8 & 1 \\ 0 & 1 & 0 \end{bmatrix} = E_1 A$$

with $E_1 = \begin{bmatrix} \frac{1}{3} & 0 & 0 \\ 0 & 1 & 0 \\ 0 & 0 & 1 \end{bmatrix}$. *Step (2), exchange the second row and the third row of* $E_1 A$,

$$E_1 A = \begin{bmatrix} 1 & 0 & 5 \\ 0 & 8 & 1 \\ 0 & 1 & 0 \end{bmatrix} \rightarrow \begin{bmatrix} 1 & 0 & 5 \\ 0 & 1 & 0 \\ 0 & 8 & 1 \end{bmatrix} = E_2 (E_1 A)$$

with $E_2 = \begin{bmatrix} 1 & 0 & 0 \\ 0 & 0 & 1 \\ 0 & 1 & 0 \end{bmatrix}$. *Step (3), replacement of the third row of* $E_2 E_1 A$,

$$E_2 E_1 A = \begin{bmatrix} 1 & 0 & 5 \\ 0 & 1 & 0 \\ 0 & 8 & 1 \end{bmatrix} \rightarrow \begin{bmatrix} 1 & 0 & 5 \\ 0 & 1 & 0 \\ 0 & 0 & 1 \end{bmatrix} = E_3 (E_2 E_1 A)$$

with $E_3 = \begin{bmatrix} 1 & 0 & 0 \\ 0 & 1 & 0 \\ 0 & -8 & 1 \end{bmatrix}$. *Lastly, Step (4), replacement of the first row of* $E_3 E_2 E_1 A$,

$$E_3 E_2 E_1 A = \begin{bmatrix} 1 & 0 & 5 \\ 0 & 1 & 0 \\ 0 & 0 & 1 \end{bmatrix} \rightarrow \begin{bmatrix} 1 & 0 & 0 \\ 0 & 1 & 0 \\ 0 & 0 & 1 \end{bmatrix} = E_4 (E_3 E_2 E_1 A)$$

with $E_4 = \begin{bmatrix} 1 & 0 & -5 \\ 0 & 1 & 0 \\ 0 & 0 & 1 \end{bmatrix}$. *We notice that* A *has been reduced to an identity matrix* I *after a number of elementary row operations, i.e.,* $E_4 E_3 E_2 E_1 A = I$. *In the following, we try to express* A *in terms of elementary matrices. To isolate* A, *we multiply both sides of*

$$E_4 E_3 E_2 E_1 A = I$$

by E_4^{-1},

$$E_4^{-1} (E_4 E_3 E_2 E_1 A) = E_4^{-1} (I)$$

which results in,

$$E_3 E_2 E_1 A = E_4^{-1}.$$

Next, we multiply both sides of the equation by E_3^{-1}, *to obtain*

$$E_3^{-1} (E_3 E_2 E_1 A) = E_3^{-1} \left(E_4^{-1} \right)$$

which is,

$$E_2 E_1 A = E_3^{-1} E_4^{-1}$$

Continuing in this fashion, we can express A as a product of elementary matrices,

$$\begin{aligned} A &= E_1^{-1}E_2^{-1}E_3^{-1}E_4^{-1} \\ &= \begin{bmatrix} \frac{1}{3} & 0 & 0 \\ 0 & 1 & 0 \\ 0 & 0 & 1 \end{bmatrix}^{-1} \begin{bmatrix} 1 & 0 & 0 \\ 0 & 0 & 1 \\ 0 & 1 & 0 \end{bmatrix}^{-1} \begin{bmatrix} 1 & 0 & 0 \\ 0 & 1 & 0 \\ 0 & -8 & 1 \end{bmatrix}^{-1} \begin{bmatrix} 1 & 0 & -5 \\ 0 & 1 & 0 \\ 0 & 0 & 1 \end{bmatrix}^{-1} \\ &= \begin{bmatrix} 3 & 0 & 0 \\ 0 & 1 & 0 \\ 0 & 0 & 1 \end{bmatrix} \begin{bmatrix} 1 & 0 & 0 \\ 0 & 0 & 1 \\ 0 & 1 & 0 \end{bmatrix} \begin{bmatrix} 1 & 0 & 0 \\ 0 & 1 & 0 \\ 0 & 8 & 1 \end{bmatrix} \begin{bmatrix} 1 & 0 & 5 \\ 0 & 1 & 0 \\ 0 & 0 & 1 \end{bmatrix} \end{aligned}$$

Theorem 12 *If A is an $n \times n$ matrix that can be row reduced to an identity matrix with k elementary operations, i.e., $E_k \cdots E_2E_1A = I$ for some elementary matrices $E_1, E_2, \ldots,$ and E_k, then A is invertible and its inverse $A^{-1} = E_k \cdots E_2E_1$. In addition, the matrix $A = (E_k \cdots E_2E_1)^{-1} = E_1^{-1}E_2^{-1} \cdots E_k^{-1}$.*

Applying this Theorem, we have the following algorithm for finding A^{-1}.

(1) Place side by side, the $n \times n$ matrix A and the identity matrix I, i.e., $[A \ I]$.

(2) Apply elementary row operations $E_1, E_2, \ldots,$ and E_k to the rectangular matrix $[A \ I]$, that is, $[E_k \cdots E_2E_1A \ \ E_k \cdots E_2E_1I]$, until A is reduced to the identity matrix.

(3) When we arrive at the form $[I \ \ E_k \cdots E_2E_1] = \left[I \ \ A^{-1}\right]$, then A^{-1} is found.

Example 13 Consider

$$A = \begin{bmatrix} 3 & 0 & 15 \\ 0 & 8 & 1 \\ 0 & 1 & 0 \end{bmatrix}$$

Find A^{-1}.

Solution *Let's place A and I side by side*

$$[A \ I] = \begin{bmatrix} 3 & 0 & 15 & 1 & 0 & 0 \\ 0 & 8 & 1 & 0 & 1 & 0 \\ 0 & 1 & 0 & 0 & 0 & 1 \end{bmatrix}$$

Next, row reduce until A is reduced to an identity matrix,

$$\begin{bmatrix} 3\,0\,15\,1\,0\,0 \\ 0\,8\ \ 1\ \ 0\,1\,0 \\ 0\,1\ \ 0\ \ 0\,0\,1 \end{bmatrix} \sim \begin{bmatrix} 1\,0\,5\,\frac{1}{3}\,0\,0 \\ 0\,8\,1\,0\,1\,0 \\ 0\,1\,0\,0\,0\,1 \end{bmatrix} \sim \begin{bmatrix} 1\,0\,5\,\frac{1}{3}\,0\,0 \\ 0\,1\,0\,0\,0\,1 \\ 0\,8\,1\,0\,1\,0 \end{bmatrix} \sim$$

$$\begin{bmatrix} 1\,0\,5\,\frac{1}{3}\,0\ \ 0 \\ 0\,1\,0\,0\,0\ \ 1 \\ 0\,0\,1\,0\,1\,-8 \end{bmatrix} \sim \begin{bmatrix} 1\,0\,0\,\frac{1}{3}\,-5\ \ 40 \\ 0\,1\,0\,0\ \ 0\ \ \ 1 \\ 0\,0\,1\,0\ \ 1\ \ -8 \end{bmatrix}$$

Thus,

$$A^{-1} = \begin{bmatrix} \frac{1}{3} & -5 & 40 \\ 0 & 0 & 1 \\ 0 & 1 & -8 \end{bmatrix}$$

3.2 Exercises

Exercise 14 Decide if each matrix A is invertible. If it is, find its inverse.

(a) $A = \begin{bmatrix} 2 & 0 \\ 0 & -3 \end{bmatrix}$ (b) $A = \begin{bmatrix} 1 & 3 \\ 3 & 4 \end{bmatrix}$ (c) $A = \begin{bmatrix} 3 & 5 \\ -2 & 2 \end{bmatrix}$

Exercise 15 Consider the equation $A\mathbf{x} = \mathbf{b}$, with $A = \begin{bmatrix} 4 & 1 \\ 2 & 3 \end{bmatrix}$. Solve each equation when $\mathbf{b}$ is given as follows.

(a) $\mathbf{b} = \begin{bmatrix} 1 \\ -1 \end{bmatrix}$ (b) $\mathbf{b} = \begin{bmatrix} 2 \\ 5 \end{bmatrix}$ (c) $\mathbf{b} = \begin{bmatrix} -3 \\ 2 \end{bmatrix}$

Exercise 16 Given elementary matrices $E_1 = \begin{bmatrix} 1 & 0 & 0 \\ 0 & 1 & 0 \\ 0 & 0 & -2 \end{bmatrix}$, $E_2 = \begin{bmatrix} 0 & 1 & 0 \\ 1 & 0 & 0 \\ 0 & 0 & 1 \end{bmatrix}$ and $E_3 = \begin{bmatrix} 1 & 0 & 0 \\ 0 & 1 & 0 \\ 5 & 0 & 1 \end{bmatrix}$. Find the products E_1A, E_2A, and E_3A for a general 3×3 matrix $A = \begin{bmatrix} a_{11} & a_{12} & a_{13} \\ a_{21} & a_{22} & a_{23} \\ a_{31} & a_{32} & a_{33} \end{bmatrix}$.

Exercise 17 Find the inverse of each matrix if it exists.

(a) $A = \begin{bmatrix} 1 & -2 & 0 \\ 0 & 1 & 3 \\ 0 & 0 & 2 \end{bmatrix}$ (b) $A = \begin{bmatrix} 1 & 0 & 0 \\ 0 & -4 & 8 \\ 2 & 0 & 2 \end{bmatrix}$

(c) $A = \begin{bmatrix} 3 & 2 & 5 \\ 2 & -1 & 8 \\ 2 & -1 & 8 \end{bmatrix}$ (d) $A = \begin{bmatrix} 1 & 2 & 3 \\ 4 & 5 & 6 \\ 7 & 8 & 9 \end{bmatrix}$

(e) $A = \begin{bmatrix} 1 & 2 & 3 & 1 \\ -1 & 0 & 5 & 1 \\ 2 & 1 & -3 & 0 \\ 0 & 2 & 5 & 2 \end{bmatrix}$

Exercise 18 Find the inverse of $A = \begin{bmatrix} 1 & 2 & 0 \\ 0 & 1 & -3 \\ 0 & 0 & 2 \end{bmatrix}$ and express A^{-1} as a product of elementary matrices.

3.3 Non-singular Transformations

Inverse of a Linear Transformation

A linear transformation T is said to be **non-singular** if there exists a linear transformation T^* such that both T^*T and TT^* are identity mappings. Otherwise T is said to be **singular**.

Let T denote a linear transformation $T(\mathbf{x}) = A\mathbf{x}$, with A being a square $n \times n$ matrix. If the matrix A is invertible, the A^{-1} matrix can define a linear transformation $T^*(\mathbf{y}) = A^{-1}\mathbf{y}$. In such a case, we have

$$T^*T(\mathbf{x}) = A^{-1}(A\mathbf{x}) = \left(A^{-1}A\right)\mathbf{x} = I_n\mathbf{x} = \mathbf{x}$$

and

$$TT^*(\mathbf{y}) = A\left(A^{-1}\mathbf{y}\right) = \left(AA^{-1}\right)\mathbf{y} = I_n\mathbf{y} = \mathbf{y}$$

which show that T^* is the inverse transformation of T. For the same reason, T is an inverse of T^*.

Example 1 Let $T : R^2 \to R^2$ be the linear transformation $T(\mathbf{x}) = A\mathbf{x} = \begin{bmatrix} 1 & -4 \\ 3 & -9 \end{bmatrix}\begin{bmatrix} x_1 \\ x_2 \end{bmatrix}$. Decide whether T is non-singular.

Solution *The transformation matrix A is invertible and its inverse*

$$A^{-1} = \frac{1}{\det A}\begin{bmatrix} -9 & 4 \\ -3 & 1 \end{bmatrix} = \frac{1}{3}\begin{bmatrix} -9 & 4 \\ -3 & 1 \end{bmatrix} = \begin{bmatrix} -3 & \frac{4}{3} \\ -1 & \frac{1}{3} \end{bmatrix}$$

Thus T^* defined by $T^*(\mathbf{x}) = A^{-1}\mathbf{x}$ is the inverse transformation of T. So T is non-singular.

Example 2 Let $T : R^3 \rightarrow R^3$ be the linear transformation $T(\mathbf{x}) = A\mathbf{x} = \begin{bmatrix} 1 & 0 & 0 \\ 0 & 6 & 0 \\ 0 & 0 & 1 \end{bmatrix} \begin{bmatrix} x_1 \\ x_2 \\ x_3 \end{bmatrix}$.

Decide whether T is non-singular.

Solution *The transformation matrix A is in fact an elementary matrix. Its inverse matrix is also an elementary matrix*

$$A^{-1} = \begin{bmatrix} 1 & 0 & 0 \\ 0 & \frac{1}{6} & 0 \\ 0 & 0 & 1 \end{bmatrix}$$

Thus T^* *defined by* $T^*(\mathbf{x}) = A^{-1}\mathbf{x}$ *is the inverse transformation of* T. *So* T *is non-singular.*

Example 3 Let $T : R^3 \rightarrow R^3$ be the linear transformation $T(\mathbf{x}) = A\mathbf{x} = \begin{bmatrix} 0 & 1 & 0 \\ 1 & 0 & 0 \\ 0 & 0 & 1 \end{bmatrix} \begin{bmatrix} x_1 \\ x_2 \\ x_3 \end{bmatrix}$.

Decide whether T is non-singular.

Solution *The transformation matrix A is an elementary matrix with its inverse matrix being the same as itself,*

$$A^{-1} = \begin{bmatrix} 0 & 1 & 0 \\ 1 & 0 & 0 \\ 0 & 0 & 1 \end{bmatrix}$$

Thus the inverse transformation of T *exists as* $T^*(\mathbf{x}) = A^{-1}\mathbf{x}$. *Hence* T *is non-singular.*

Example 4 Let $T : R^3 \rightarrow R^3$ be the linear transformation $T(\mathbf{x}) = A\mathbf{x} = \begin{bmatrix} 1 & 0 & 0 \\ 0 & 1 & 0 \\ 3 & 0 & 1 \end{bmatrix} \begin{bmatrix} x_1 \\ x_2 \\ x_3 \end{bmatrix}$.

Decide whether T is non-singular.

Solution *The transformation matrix A is an elementary matrix and its inverse matrix*

$$A^{-1} = \begin{bmatrix} 1 & 0 & 0 \\ 0 & 1 & 0 \\ -3 & 0 & 1 \end{bmatrix}$$

defines the inverse transformation as $T^*(\mathbf{x}) = A^{-1}\mathbf{x}$. *So* T *is non-singular.*

Characteristics of Non-singular Transformations

Let T denote a linear transformation $T(\mathbf{x}) = A\mathbf{x}$, with A being a square $n \times n$ matrix. The following statements can be proved to be equivalent. All the following statements are either all true or all false.

(1) The linear transformation $T(\mathbf{x}) = A\mathbf{x}$ is non-singular.
(2) The matrix A is invertible.
(3) There exists an $n \times n$ matrix C such that $CA = I_n$.
(4) There exists an $n \times n$ matrix D such that $AD = I_n$.
(5) The matrix A is row equivalent to an identity matrix I_n.
(6) Each column of A is a pivot column.
(7) The matrix equation $A\mathbf{x} = \mathbf{b}$ has one and only one solution for each $\mathbf{b}$ in R^n.
(8) The linear transformation $T(\mathbf{x}) = A\mathbf{x}$ is onto.
(9) The columns of A span R^n.
(10) The matrix equation $A\mathbf{x} = \mathbf{0}$ has only the trivial solution.
(11) The columns of A form a linearly independent set.
(12) The linear transformation $T(\mathbf{x}) = A\mathbf{x}$ is one-to-one.

Example 5 Let $T : R^3 \to R^3$ be the linear transformation $T(\mathbf{x}) = A\mathbf{x} = \begin{bmatrix} 1 & 3 & 5 \\ 2 & 6 & 5 \\ 3 & 9 & 5 \end{bmatrix} \begin{bmatrix} x_1 \\ x_2 \\ x_3 \end{bmatrix}$.

Decide whether T is non-singular.

Solution *Let us row reduce the transformation matrix A,*

$$\begin{bmatrix} 1 & 3 & 5 \\ 2 & 6 & 5 \\ 3 & 9 & 5 \end{bmatrix} \sim \begin{bmatrix} 1 & 3 & 5 \\ 0 & 0 & -5 \\ 0 & 0 & -10 \end{bmatrix} \sim \begin{bmatrix} 1 & 3 & 0 \\ 0 & 0 & 1 \\ 0 & 0 & 0 \end{bmatrix}$$

The matrix A does not have an inverse since it is not row equivalent an identity matrix. Therefore, $T(\mathbf{x}) = A\mathbf{x}$ is a singular transformation.

Remark 6 When a linear transformation is defined by an $n \times n$ matrix A whose columns span R^n, the transformation has an inverse and hence it is non-singular. Otherwise it is singular.

3.3 Exercises

Exercise 7 Decide if each mapping $T(\mathbf{x}) = A\mathbf{x}$ is non-singular. If it is non-singular, find its inverse.

(a) $A = \begin{bmatrix} 2 & 0 \\ 0 & -1 \end{bmatrix}$ (b) $A = \begin{bmatrix} 1 & 3 \\ 0 & 4 \end{bmatrix}$ (c) $A = \begin{bmatrix} -1 & 2 \\ 3 & -6 \end{bmatrix}$ (d) $A = \begin{bmatrix} -2 & 5 \\ 1 & -4 \end{bmatrix}$

Exercise 8 Let $T : R^3 \to R^3$ be the linear transformation $T(\mathbf{x}) = A\mathbf{x} = \begin{bmatrix} 1 & 0 & 2 \\ 0 & 3 & 8 \\ 0 & 0 & -1 \end{bmatrix} \begin{bmatrix} x_1 \\ x_2 \\ x_3 \end{bmatrix}$.

Decide whether T is non-singular.

Exercise 9 Let $S : R^3 \to R^3$ be the linear transformation $S(\mathbf{x}) = A\mathbf{x} = \begin{bmatrix} 3 & 2 & 7 \\ 0 & 1 & 6 \\ 0 & 0 & 5 \end{bmatrix} \begin{bmatrix} x_1 \\ x_2 \\ x_3 \end{bmatrix}$.

Decide whether $T(\mathbf{x}) = A^T\mathbf{x} = \begin{bmatrix} 3 & 0 & 0 \\ 2 & 1 & 0 \\ 7 & 6 & 5 \end{bmatrix} \begin{bmatrix} x_1 \\ x_2 \\ x_3 \end{bmatrix}$ is non-singular.

Exercise 10 Let $T : R^3 \to R^3$ be the linear transformation $T(\mathbf{x}) = A\mathbf{x} = \begin{bmatrix} 1 & 3 & 1 \\ 2 & 6 & 2 \\ 0 & 0 & 1 \end{bmatrix} \begin{bmatrix} x_1 \\ x_2 \\ x_3 \end{bmatrix}$.

Decide whether T is non-singular.

Exercise 11 Let $T : R^3 \to R^3$ be the linear transformation $T(\mathbf{x}) = A\mathbf{x} = \begin{bmatrix} 1 & 0 & 2 \\ 3 & 0 & 5 \\ 0 & 0 & 1 \end{bmatrix} \begin{bmatrix} x_1 \\ x_2 \\ x_3 \end{bmatrix}$.

Decide whether T is non-singular.

Exercise 12 For each given A, determine (a) if the columns of A span R^3; (b) if the mapping $T(\mathbf{x}) = A\mathbf{x}$ is invertible; (c) if the mapping T is non-singular.

$$\text{(a) } A = \begin{bmatrix} 1 & 0 & 0 \\ 0 & 1 & 0 \\ 0 & 0 & 1 \end{bmatrix} \qquad \text{(b) } A = \begin{bmatrix} 1 & 0 & 0 \\ 0 & 1 & 0 \\ 0 & 0 & 0 \end{bmatrix}$$

Exercise 13 Find the inverse transformation of $T(\mathbf{x}) = A\mathbf{x} = \begin{bmatrix} 1 & -1 & 1 \\ 0 & -1 & 2 \\ 2 & -2 & 1 \end{bmatrix} \begin{bmatrix} x_1 \\ x_2 \\ x_3 \end{bmatrix}$.

3.4 Vector Spaces

Properties of a Vector Space

A **vector space** V is a set of vectors in R^n satisfying three properties:

1. the $\mathbf{0}$ vector is in V.
2. the vector $\mathbf{u} + \mathbf{v}$ is in V, for any $\mathbf{u}$ and $\mathbf{v}$ in V.
3. the vector $c\mathbf{u}$ is in V, for any scalar c and any $\mathbf{u}$ in V.

A vector space V is also called a subspace of R^n.

Example 1 Let $V = Span\left\{\begin{bmatrix}1\\0\end{bmatrix}\right\}$. We can verify that V satisfies the three properties of a vector space. (1) the $\mathbf{0} = \begin{bmatrix}0\\0\end{bmatrix}$ vector is in V; (2) the vector $\mathbf{u}+\mathbf{v}$ is in V, for any $\mathbf{u}$ and $\mathbf{v}$ in V. Let's suppose $\mathbf{u}$ and $\mathbf{v}$ are both in V, i.e.,

$$\mathbf{u}{=}s\begin{bmatrix}1\\0\end{bmatrix}$$

$$\mathbf{v} = t\begin{bmatrix}1\\0\end{bmatrix}$$

then

$$\mathbf{u}+\mathbf{v} = (s+t)\begin{bmatrix}1\\0\end{bmatrix}$$

which is in V. Next we check on the third property (3) the vector $c\mathbf{u}$ is in V, for any scalar c and any $\mathbf{u}$ in V. We suppose $\mathbf{u}$ is in V, i.e.,

$$\mathbf{u}{=}s\begin{bmatrix}1\\0\end{bmatrix}$$

then the scalar multiple of $\mathbf{u}$

$$c\mathbf{u} = (cs)\begin{bmatrix}1\\0\end{bmatrix}$$

is also in V. Therefore V is a subspace of R^2.

Example 2 Let $V = Span\left\{\begin{bmatrix}1\\0\end{bmatrix}, \begin{bmatrix}0\\1\end{bmatrix}\right\}$. We can verify that V satisfies the three properties of a vector space. (1). the $\mathbf{0} = \begin{bmatrix}0\\0\end{bmatrix}$ vector is in V; (2) the vector $\mathbf{u}+\mathbf{v}$ is in V, for any $\mathbf{u}$ and $\mathbf{v}$ in V. Let's suppose $\mathbf{u}$ and $\mathbf{v}$ are both in V, i.e.,

$$\mathbf{u}{=}\begin{bmatrix}u_1\\u_2\end{bmatrix} = u_1\begin{bmatrix}1\\0\end{bmatrix} + u_2\begin{bmatrix}0\\1\end{bmatrix}$$

$$\mathbf{v}{=}\begin{bmatrix}v_1\\v_2\end{bmatrix} = v_1\begin{bmatrix}1\\0\end{bmatrix} + v_2\begin{bmatrix}0\\1\end{bmatrix}$$

then

$$\mathbf{u}+\mathbf{v} = (u_1+v_1)\begin{bmatrix}1\\0\end{bmatrix} + (u_2+v_2)\begin{bmatrix}0\\1\end{bmatrix}$$

which is in V. Next we check on the third property (3) the vector $c\mathbf{u}$ is in V, for any scalar c and any $\mathbf{u}$ in V. We suppose $\mathbf{u}$ is in V, i.e.,

$$\mathbf{u} = \begin{bmatrix} u_1 \\ u_2 \end{bmatrix} = u_1 \begin{bmatrix} 1 \\ 0 \end{bmatrix} + u_2 \begin{bmatrix} 0 \\ 1 \end{bmatrix}$$

then the scalar multiple of $\mathbf{u}$

$$c\mathbf{u} = \begin{bmatrix} cu_1 \\ cu_2 \end{bmatrix} = cu_1 \begin{bmatrix} 1 \\ 0 \end{bmatrix} + cu_2 \begin{bmatrix} 0 \\ 1 \end{bmatrix}$$

is also in V. Therefore V is a subspace of R^2. As a matter of fact, this subspace V is the same as R^2.

Example 3 Let $W = Span\{\mathbf{w}_1, \mathbf{w}_2\}$, where $\mathbf{w}_1$ and $\mathbf{w}_2$ are independent vectors in R^n. Let's decide if the set W satisfies the three properties of a vector space. (1) we know that the $\mathbf{0}$ vector in R^n is in W; (2) the vector $\mathbf{u} + \mathbf{v}$ is in W, for any $\mathbf{u}$ and $\mathbf{v}$ in W. Let's suppose $\mathbf{u}$ and $\mathbf{v}$ are both in W, i.e.,

$$\mathbf{u} = u_1\mathbf{w}_1 + u_2\mathbf{w}_2$$
$$\mathbf{v} = v_1\mathbf{w}_1 + v_2\mathbf{w}_2$$

then

$$\mathbf{u} + \mathbf{v} = (u_1 + v_1)\,\mathbf{w}_1 + (u_2 + v_2)\,\mathbf{w}_2$$

which is shown to be in W. Next we check on the third property (3) the vector $c\mathbf{u}$ is in W, for any scalar c and any $\mathbf{u}$ in W. We suppose $\mathbf{u}$ is in W, i.e.,

$$\mathbf{u} = u_1\mathbf{w}_1 + u_2\mathbf{w}_2$$

then the scalar multiple of $\mathbf{u}$

$$c\mathbf{u} = cu_1\mathbf{w}_1 + cu_2\mathbf{w}_2$$

is also in W. Therefore W is a subspace of R^n.

Remark 4 In the above example, the subspace W is the same as R^n only when $n = 2$.

Remark 5 Let $W = Span\left\{\mathbf{w}_1, \mathbf{w}_2, \ldots, \mathbf{w}_q\right\}$, with $\mathbf{w}_j,\ j = 1, 2, \ldots, q$ in R^n. We can similarly show that W satisfies the three properties of a subspace of R^n. This W is said to be a subspace spanned by $\mathbf{w}_1, \mathbf{w}_2, \ldots, \mathbf{w}_q$.

The following statements are always true for a subspace V of R^n, with the vector addition and scalar multiplication. In the following, c and d are scalars, and $\mathbf{u}$, $\mathbf{v}$, $\mathbf{w}$ are any arbitrary vectors in V.

1. $\mathbf{u}+\mathbf{v}=\mathbf{v}+\mathbf{u}$
2. $(\mathbf{u}+\mathbf{v})+\mathbf{w}=\mathbf{u}+(\mathbf{v}+\mathbf{w})$
3. There is a $\mathbf{0}$ vector in V such that $\mathbf{0}+\mathbf{u}=\mathbf{u}$
4. For any $\mathbf{u}$ in V, there is a vector $-\mathbf{u}$ such that $\mathbf{u}+(-\mathbf{u})=\mathbf{0}$
5. $c(\mathbf{u}+\mathbf{v})=c\mathbf{u}+c\mathbf{v}$
6. $(c+d)\mathbf{u}=c\mathbf{u}+d\mathbf{u}$
7. $1\mathbf{u}=\mathbf{u}$
8. $c(d\mathbf{u})=(cd)\mathbf{u}$.

Basis for a Vector Space

A **basis** for a subspace V of R^n is a linearly independent set in V that generates V. In order for a set of vectors $\{\mathbf{v}_1, \mathbf{v}_2, \ldots, \mathbf{v}_k\}$ to be a basis for a subspace V of R^n, the set must satisfy two conditions: (i) $V = Span\{\mathbf{v}_1, \mathbf{v}_2, \ldots, \mathbf{v}_k\}$ and (ii) $\{\mathbf{v}_1, \mathbf{v}_2, \ldots, \mathbf{v}_k\}$ is linearly independent.

Example 6 Consider $V = Span\left\{\begin{bmatrix}1\\0\\0\end{bmatrix}, \begin{bmatrix}0\\1\\0\end{bmatrix}, \begin{bmatrix}-1\\2\\0\end{bmatrix}\right\}$. We notice that $\begin{bmatrix}-1\\2\\0\end{bmatrix}$ can be generated by the vectors $\begin{bmatrix}1\\0\\0\end{bmatrix}$ and $\begin{bmatrix}0\\1\\0\end{bmatrix}$. So a basis for the subspace V can be the set $\left\{\begin{bmatrix}1\\0\\0\end{bmatrix}, \begin{bmatrix}0\\1\\0\end{bmatrix}\right\}$ or the set $\left\{\begin{bmatrix}1\\0\\0\end{bmatrix}, \begin{bmatrix}-1\\2\\0\end{bmatrix}\right\}$, or the set $\left\{\begin{bmatrix}0\\1\\0\end{bmatrix}, \begin{bmatrix}-1\\2\\0\end{bmatrix}\right\}$. Each set of two independent vectors can span the same vector space V.

Column Space of a Matrix

If we represent an $m \times n$ matrix A using its columns, i.e., $A = [\mathbf{a}_1\ \mathbf{a}_2\ \cdots\ \mathbf{a}_n]$, we can study the subspace spanned by the columns, $Span\{\mathbf{a}_1, \mathbf{a}_2, \ldots, \mathbf{a}_n\}$, which is denoted by $Col\ A$, and is called the **column space** of A. That is,

$$Col\ A = Span\{\mathbf{a}_1, \mathbf{a}_2, \ldots, \mathbf{a}_n\}$$

Note that each column of A is a vector in R^m. So $Col\ A$ is a subspace of R^m.

Example 7 Let $A = \begin{bmatrix}1 & 1\\0 & 2\\0 & 0\end{bmatrix}$. Decide if $\mathbf{b} = \begin{bmatrix}1\\2\\3\end{bmatrix}$ is in the column space of A.

Solution *The matrix A has two columns and the column space is,*

$$Col\ A = Span\left\{\begin{bmatrix}1\\0\\0\end{bmatrix},\begin{bmatrix}1\\2\\0\end{bmatrix}\right\}$$

To decide if **b** *is in* $Col\ A$*, we need to check if* **b** *can be generated by the columns of* A*. The vector* **b** *is in* $Col\ A$ *if and only if the following system has a solution,*

$$x_1\begin{bmatrix}1\\0\\0\end{bmatrix}+x_2\begin{bmatrix}1\\2\\0\end{bmatrix}=\begin{bmatrix}1\\2\\3\end{bmatrix}$$

But the augmented matrix,

$$\begin{bmatrix}1&1&1\\0&2&2\\0&0&3\end{bmatrix}$$

indicates inconsistency. Therefore **b** *is not in* $Col\ A$.

Example 8 Let $A=\begin{bmatrix}1&0&-1&2\\0&1&2&0\\0&0&0&0\end{bmatrix}$. Find a basis for the $Col\ A$.

Solution *The matrix A has 4 columns, and*

$$Col\ A = Span\left\{\begin{bmatrix}1\\0\\0\end{bmatrix},\begin{bmatrix}0\\1\\0\end{bmatrix},\begin{bmatrix}-1\\2\\0\end{bmatrix},\begin{bmatrix}2\\0\\0\end{bmatrix}\right\}$$

We notice that the first two columns of A are independent. But the third and fourth columns are dependent on the first two columns,

$$\mathbf{a}_3 = -\mathbf{a}_1 + 2\mathbf{a}_2$$
$$\mathbf{a}_4 = 2\mathbf{a}_1$$

The $Col\ A$ *has all the vectors generated by the 4 columns, i.e.,*

$$\mathbf{v} = c_1\mathbf{a}_1 + c_2\mathbf{a}_2 + c_3\mathbf{a}_3 + c_4\mathbf{a}_4$$

which can be equivalently written as,

$$\begin{aligned}\mathbf{v} &= c_1\mathbf{a}_1 + c_2\mathbf{a}_2 + c_3\left(-\mathbf{a}_1 + 2\mathbf{a}_2\right) + c_4\left(2\mathbf{a}_1\right)\\ &= \left(c_1 - c_3 + 2c_4\right)\mathbf{a}_1 + \left(c_2 + 2c_3\right)\mathbf{a}_2\end{aligned}$$

This means that the Col A can be generated by the two pivot columns $\mathbf{a}_1$ *and* $\mathbf{a}_2$*.So a basis for Col A is* $\{\mathbf{a}_1, \mathbf{a}_2\} = \left\{ \begin{bmatrix} 1 \\ 0 \\ 0 \end{bmatrix}, \begin{bmatrix} 0 \\ 1 \\ 0 \end{bmatrix} \right\}$.

Example 9 Let $A = \begin{bmatrix} 1 & -2 & 0 & -1 & 3 \\ 0 & 0 & 1 & 2 & -2 \\ 0 & 0 & 0 & 0 & 0 \end{bmatrix}$. Find a basis for $Col\ A$.

Solution *The matrix has 5 columns* $A = [\mathbf{a}_1\ \mathbf{a}_2\ \mathbf{a}_3\ \mathbf{a}_4\ \mathbf{a}_5]$. *The two pivot columns are* $\mathbf{a}_1$ *and* $\mathbf{a}_3$. *The other three columns are linear combinations of the two pivot columns,*

$$\mathbf{a}_2 = -2\mathbf{a}_1$$
$$\mathbf{a}_4 = -\mathbf{a}_1 + 2\mathbf{a}_3$$
$$\mathbf{a}_5 = 3\mathbf{a}_1 - 2\mathbf{a}_3$$

The Col A is the space spanned by the 5 columns, and it has all the vectors of the form

$$\mathbf{v} = c_1\mathbf{a}_1 + c_2\mathbf{a}_2 + c_3\mathbf{a}_3 + c_4\mathbf{a}_4 + c_5\mathbf{a}_5$$

which can be equivalently written as,

$$\begin{aligned} \mathbf{v} &= c_1\mathbf{a}_1 + c_2(-2\mathbf{a}_1) + c_3\mathbf{a}_3 + c_4(-\mathbf{a}_1 + 2\mathbf{a}_3) + c_5(3\mathbf{a}_1 - 2\mathbf{a}_3) \\ &= (c_1 - 2c_2 - c_4 + 3c_5)\,\mathbf{a}_1 + (c_3 + 2c_4 - 2c_5)\,\mathbf{a}_3 \end{aligned}$$

This means that the Col A can be generated by the two pivot columns $\mathbf{a}_1$ *and* $\mathbf{a}_3$. *A basis for Col A is* $\{\mathbf{a}_1, \mathbf{a}_3\}$, *i.e.,* $\left\{ \begin{bmatrix} 1 \\ 0 \\ 0 \end{bmatrix}, \begin{bmatrix} 0 \\ 1 \\ 0 \end{bmatrix} \right\}$.

Example 10 Let $B = \begin{bmatrix} 1 & -2 & 2 & 3 & -1 \\ 0 & 0 & 5 & 10 & -10 \\ 0 & 0 & 1 & 2 & -2 \end{bmatrix}$. Find a basis for $Col\ B$.

Solution *We apply elementary row operations to reduce the matrix B to its reduced row echelon form*

$$A = \begin{bmatrix} 1 & -2 & 0 & -1 & 3 \\ 0 & 0 & 1 & 2 & -2 \\ 0 & 0 & 0 & 0 & 0 \end{bmatrix}$$

which shows that the pivot columns of B are columns 1 *and* 3. *The row operations will not affect the linear dependence relationship among columns of B. Therefore, a basis for $Col\ B$ is* $\left\{\begin{bmatrix}1\\0\\0\end{bmatrix}, \begin{bmatrix}2\\5\\1\end{bmatrix}\right\}$.

Remark 11 To provide a basis for $Col\ B$ in the above example, we use the pivot columns of B, instead of the columns 1 and 3 of the reduced row echelon form A. The columns of A can't generate $Col\ B$ because all columns of A have 0 in their last entries.

Theorem 12 *The pivot columns of a matrix form a basis for its column space.*

Null Space of a Matrix

The **null space** of a matrix A, denoted by $Nul\ A$, is the set of all solutions of $A\mathbf{x} = \mathbf{0}$. If A is an $m \times n$ matrix, the null space of A is in fact a subspace of R^n.

Example 13 Let $B = \begin{bmatrix}1 & -2 & 2 & 3 & -1\\0 & 0 & 5 & 10 & -10\\0 & 0 & 1 & 2 & -2\end{bmatrix}$. Find a basis for $Nul\ B$.

Solution *This matrix B can be row reduced to the matrix A,*

$$A = \begin{bmatrix}1 & -2 & 0 & -1 & 3\\0 & 0 & 1 & 2 & -2\\0 & 0 & 0 & 0 & 0\end{bmatrix}$$

The general solution of $A\mathbf{x} = \mathbf{0}$ has 3 *free variables, and*

$$\mathbf{x} = \begin{bmatrix}x_1\\x_2\\x_3\\x_4\\x_5\end{bmatrix} = x_2\begin{bmatrix}2\\1\\0\\0\\0\end{bmatrix} + x_4\begin{bmatrix}1\\0\\-2\\1\\0\end{bmatrix} + x_5\begin{bmatrix}-3\\0\\2\\0\\1\end{bmatrix}$$

So *Nul* $A = Span \left\{ \begin{bmatrix} 2 \\ 1 \\ 0 \\ 0 \\ 0 \end{bmatrix}, \begin{bmatrix} 1 \\ 0 \\ -2 \\ 1 \\ 0 \end{bmatrix}, \begin{bmatrix} -3 \\ 0 \\ 2 \\ 0 \\ 1 \end{bmatrix} \right\}$, and a basis for *Nul* A is $\left\{ \begin{bmatrix} 2 \\ 1 \\ 0 \\ 0 \\ 0 \end{bmatrix}, \begin{bmatrix} 1 \\ 0 \\ -2 \\ 1 \\ 0 \end{bmatrix}, \begin{bmatrix} -3 \\ 0 \\ 2 \\ 0 \\ 1 \end{bmatrix} \right\}$.

Theorem 14 *(i) The null space of an $m \times n$ matrix A is a subspace of R^n. (ii) The column space of an $m \times n$ matrix A is a subspace of R^m.*

3.4 Exercises

Exercise 15 Let $V = Span \left\{ \begin{bmatrix} 2 \\ 0 \end{bmatrix} \right\}$. Is V a subspace of R^2? Why or why not?

Exercise 16 Let $V = Span \left\{ \begin{bmatrix} 0 \\ 5 \end{bmatrix} \right\}$. Is V a subspace of R^2? Why or why not?

Exercise 17 Let V be the set of vectors of the form $\begin{bmatrix} a \\ 0 \end{bmatrix}$, $a > 0$. Is V a subspace of R^2? Why or why not?

Exercise 18 Let V be the set of vectors of the form $\begin{bmatrix} a \\ 0 \end{bmatrix}$, $a \in R$. Is V a subspace of R^2? Why or why not?

Exercise 19 Let V be the set of vectors of the form $\begin{bmatrix} a \\ 2a \end{bmatrix}$, $a \in R$. Is V a subspace of R^2? Why or why not?

Exercise 20 Decide if $\mathbf{b} = \begin{bmatrix} 1 \\ 1 \\ 2 \end{bmatrix}$ is in the column space of $A = \begin{bmatrix} 1 & 2 \\ 0 & 1 \\ 1 & 1 \end{bmatrix}$.

Exercise 21 Decide if $\mathbf{w} = \begin{bmatrix} 1 \\ 1 \\ 2 \\ 1 \end{bmatrix}$ is in the null space of $A = \begin{bmatrix} 1 & 0 & -1 & 1 \\ 0 & 1 & 2 & -2 \\ 0 & 0 & 0 & 0 \end{bmatrix}$.

Exercise 22 Find a basis for the column space of $A = \begin{bmatrix} 1 & 0 & 2 \\ 0 & 1 & 0 \\ 0 & 1 & 1 \end{bmatrix}$.

Exercise 23 Find a basis for the column space of $A = \begin{bmatrix} 1 & 0 & -3 & 1 \\ 0 & 2 & 2 & -2 \\ 0 & 0 & 0 & 8 \end{bmatrix}$.

Exercise 24 Find a basis for the null space of $A = \begin{bmatrix} 1 & 0 & 2 \\ 0 & 1 & 0 \\ 0 & 1 & 1 \end{bmatrix}$.

Exercise 25 Find a basis for the null space of $A = \begin{bmatrix} 1 & 0 & -3 & 1 \\ 0 & 2 & 2 & -2 \\ 0 & 0 & 0 & 8 \end{bmatrix}$.

3.5 Dimension and Rank

Standard Basis and Other Bases for R^n

The **standard basis** for R^n is $\{\mathbf{e}_1, \mathbf{e}_2, \ldots, \mathbf{e}_n\}$, where

$$\mathbf{e}_1 = \begin{bmatrix} 1 \\ 0 \\ 0 \\ \vdots \\ 0 \end{bmatrix}, \ \mathbf{e}_2 = \begin{bmatrix} 0 \\ 1 \\ 0 \\ \vdots \\ 0 \end{bmatrix}, \ \ldots, \ \mathbf{e}_n = \begin{bmatrix} 0 \\ 0 \\ \vdots \\ 0 \\ 1 \end{bmatrix}.$$

Can we use a different set of independent vectors to generate R^n? We know the answer is yes. For example, the set $\{\mathbf{u}_1, \mathbf{u}_2, \ldots, \mathbf{u}_n\}$, where

$$\mathbf{u}_1 = \begin{bmatrix} 1 \\ 0 \\ 0 \\ \vdots \\ 0 \end{bmatrix}, \ \mathbf{u}_2 = \begin{bmatrix} 1 \\ 1 \\ 0 \\ \vdots \\ 0 \end{bmatrix}, \ \ldots, \ \mathbf{u}_n = \begin{bmatrix} 1 \\ 1 \\ \vdots \\ 1 \\ 1 \end{bmatrix},$$

can generate R^n as well. In fact, any set of n independent vectors in R^n can span R^n. There are numerous possibilities for the bases of R^n.

Example 1 The standard basis for R^3 is the set $\{\mathbf{e}_1, \mathbf{e}_2, \mathbf{e}_3\}$, where

$$\mathbf{e}_1 = \begin{bmatrix} 1 \\ 0 \\ 0 \end{bmatrix}, \ \mathbf{e}_2 = \begin{bmatrix} 0 \\ 1 \\ 0 \end{bmatrix}, \ \mathbf{e}_3 = \begin{bmatrix} 0 \\ 0 \\ 1 \end{bmatrix}.$$

Another possible basis for R^3 is the set $\{\mathbf{v}_1, \mathbf{v}_2, \mathbf{v}_3\}$, where

$$\mathbf{v}_1 = \begin{bmatrix} 5 \\ 0 \\ 0 \end{bmatrix}, \ \mathbf{v}_2 = \begin{bmatrix} 2 \\ 2 \\ 0 \end{bmatrix}, \ \mathbf{v}_3 = \begin{bmatrix} 3 \\ 0 \\ 8 \end{bmatrix}.$$

Note that the following matrix with $\mathbf{v}_1$, $\mathbf{v}_2$, and $\mathbf{v}_3$ as columns,

$$\begin{bmatrix} 5 & 2 & 3 \\ 0 & 2 & 0 \\ 0 & 0 & 8 \end{bmatrix}$$

showing 3 pivot columns. So the set $\{\mathbf{v}_1, \mathbf{v}_2, \mathbf{v}_3\}$ is a linearly independent set that span R^3.

Once a basis is chosen for R^n, any vector in R^n can be represented as a linear combination of the basis vectors.

Dimension and Coordinate Relative to a Basis

Consider a subspace V of R^n. A **basis** for a vector space V of R^n is a linearly independent set in V that generates V.

Example 2 Consider $V = Span\left\{\begin{bmatrix} 1 \\ 0 \\ 0 \end{bmatrix}, \begin{bmatrix} 0 \\ 1 \\ 0 \end{bmatrix}, \begin{bmatrix} -1 \\ 2 \\ 0 \end{bmatrix}\right\}$. We notice that $\begin{bmatrix} -1 \\ 2 \\ 0 \end{bmatrix}$ can be generated by the vectors $\begin{bmatrix} 1 \\ 0 \\ 0 \end{bmatrix}$ and $\begin{bmatrix} 0 \\ 1 \\ 0 \end{bmatrix}$. So a basis for the subspace V can be the set $\left\{\begin{bmatrix} 1 \\ 0 \\ 0 \end{bmatrix}, \begin{bmatrix} 0 \\ 1 \\ 0 \end{bmatrix}\right\}$ or the set $\left\{\begin{bmatrix} 1 \\ 0 \\ 0 \end{bmatrix}, \begin{bmatrix} -1 \\ 2 \\ 0 \end{bmatrix}\right\}$, or the set $\left\{\begin{bmatrix} 0 \\ 1 \\ 0 \end{bmatrix}, \begin{bmatrix} -1 \\ 2 \\ 0 \end{bmatrix}\right\}$. Each set of these two independent vectors can span the same vector space V.

Let's denote a basis for V as $\mathcal{B} = \{\mathbf{v}_1, \mathbf{v}_2, \ldots, \mathbf{v}_k\}$. The number of vectors in any basis for V is the **dimension** of the subspace V in R^n, denoted by dim V. By convention, dim $\{\mathbf{0}\} = 0$. We know that any vector $\mathbf{v}$ in V can be generated in terms of the basis vectors, i.e., $\mathbf{v} = c_1\mathbf{v}_1 + c_2\mathbf{v}_2 + \cdots + c_k\mathbf{v}_k$. The weights $c_1, c_2, \ldots, c_k$ are called the coordinates of $\mathbf{v}$ relative to the basis $\mathcal{B}$. The **coordinate vector of v relative to $\mathcal{B}$** which is denoted by $[\mathbf{v}]_{\mathcal{B}}$ is defined as,

$$[\mathbf{v}]_{\mathcal{B}} = \begin{bmatrix} c_1 \\ c_2 \\ \vdots \\ c_k \end{bmatrix}.$$

Theorem 3 *Let $\mathcal{B} = \{\mathbf{v}_1, \mathbf{v}_2, \ldots, \mathbf{v}_k\}$ be a basis for a subspace V of R^n. Each vector $\mathbf{v}$ in V can be uniquely represented as a linear combination of the basis vectors. That is, the coordinate vector of $\mathbf{v}$ relative to $\mathcal{B}$ is unique.*

Proof Let's suppose there exists a vector $\mathbf{u}$ in the subspace V, for which

$$\mathbf{u} = s_1\mathbf{v}_1 + s_2\mathbf{v}_2 + \cdots + s_k\mathbf{v}_k$$

and

$$\mathbf{u} = t_1\mathbf{v}_1 + t_2\mathbf{v}_2 + \cdots + t_k\mathbf{v}_k$$

Subtracting the two equations, we obtain

$$\mathbf{0} = (s_1 - t_1)\,\mathbf{v}_1 + (s_2 - t_2)\,\mathbf{v}_2 + \cdots + (s_k - t_k)\,\mathbf{v}_k$$

Since the basis vectors are independent, the homogeneous system

$$\mathbf{0} = x_1\mathbf{v}_1 + x_2\mathbf{v}_2 + \cdots + x_k\mathbf{v}_k$$

has only the trivial solution $x_1 = 0$, $x_2 = 0$, …, $x_k = 0$. Therefore, $s_1 = t_1$, $s_2 = t_2$, …, $s_k = t_k$. So each vector $\mathbf{v}$ in V is uniquely represented as a linear combination of the basis vectors. ∎

Example 4 Let V be a subspace in R^3 with the basis $\mathcal{B} = \{\mathbf{e}_1, \mathbf{e}_2\}$. (a) Decide if $\mathbf{u} = \begin{bmatrix} 5 \\ 2 \\ 0 \end{bmatrix}$ is in V. (b) If $\mathbf{u}$ is in V, find the coordinate vector of $\mathbf{u}$ relative to $\mathcal{B}$.

Solution *Since* $\mathbf{u} = 5\mathbf{e}_1 + 2\mathbf{e}_2$, *i.e.,*

$$\mathbf{u} = 5\begin{bmatrix} 1 \\ 0 \\ 0 \end{bmatrix} + 2\begin{bmatrix} 0 \\ 1 \\ 0 \end{bmatrix}$$

So the vector $\mathbf{u}$ *is in the subspace* V. *The coordinator vector of* $\mathbf{u}$ *relative to* $\mathcal{B}$ *is*

$$[\mathbf{u}]_{\mathcal{B}} = \begin{bmatrix} 5 \\ 2 \end{bmatrix}$$

Example 5 Let V be a subspace in R^3 with the basis $\widetilde{\mathcal{B}} = \{\mathbf{b}_1, \mathbf{b}_2\}$, where $\mathbf{b}_1 = \begin{bmatrix} 1 \\ 0 \\ 0 \end{bmatrix}$ and $\mathbf{b}_2 = \begin{bmatrix} 1 \\ 1 \\ 0 \end{bmatrix}$. (a) Decide if $\mathbf{u} = \begin{bmatrix} 5 \\ 2 \\ 0 \end{bmatrix}$ is in V. (b) If $\mathbf{u}$ is in V, find the coordinate vector of $\mathbf{u}$ relative to $\widetilde{\mathcal{B}}$.

Solution *Let's decide if it is possible to express* $\mathbf{u}$ *as a linear combination of the basis vectors* $\mathbf{b}_1$ *and* $\mathbf{b}_2$*. That is, to consider* $\mathbf{u} = x_1\mathbf{b}_1 + x_2\mathbf{b}_2$, *i.e.,*

$$\begin{bmatrix} 5 \\ 2 \\ 0 \end{bmatrix} = x_1 \begin{bmatrix} 1 \\ 0 \\ 0 \end{bmatrix} + x_2 \begin{bmatrix} 1 \\ 1 \\ 0 \end{bmatrix}$$

Since the above system is consistent, the vector $\mathbf{u}$ *is in the subspace generated by* $\mathbf{b}_1$ *and* $\mathbf{b}_2$*. The coordinator vector of* $\mathbf{u}$ *relative to the basis* $\widetilde{\mathcal{B}}$ *is*

$$[\mathbf{u}]_{\widetilde{\mathcal{B}}} = \begin{bmatrix} x_1 \\ x_2 \end{bmatrix} = \begin{bmatrix} 3 \\ 2 \end{bmatrix}$$

Remark 6 The subspace V in R^3 is in fact the same vector space in the previous two examples. That is, two different bases $\mathcal{B}$ and $\widetilde{\mathcal{B}}$ span the same subspace. The dimension of V is 2. For the same vector $\mathbf{u}$, its coordinate vector relative to $\widetilde{\mathcal{B}}$ is different than that relative to $\mathcal{B}$.

Example 7 Consider R^3 with the standard basis $\mathcal{B} = \{\mathbf{e}_1, \mathbf{e}_2, \mathbf{e}_3\}$. Find the coordinate vector of $\mathbf{u} = \begin{bmatrix} 5 \\ 2 \\ 0 \end{bmatrix}$ relative to $\mathcal{B}$.

Solution *Since* $\mathbf{u} = 5\mathbf{e}_1 + 2\mathbf{e}_2 + 0\mathbf{e}_3$, *i.e.,*

$$\mathbf{u} = 5 \begin{bmatrix} 1 \\ 0 \\ 0 \end{bmatrix} + 2 \begin{bmatrix} 0 \\ 1 \\ 0 \end{bmatrix} + 0 \begin{bmatrix} 0 \\ 0 \\ 1 \end{bmatrix}$$

So the coordinator vector of $\mathbf{u}$ *relative to* $\mathcal{B}$ *is*

$$[\mathbf{u}]_{\mathcal{B}} = \begin{bmatrix} 5 \\ 2 \\ 0 \end{bmatrix}$$

Example 8 Consider R^3 with the basis $\widetilde{\mathcal{B}} = \{\mathbf{b}_1, \mathbf{b}_2, \mathbf{b}_3\}$, where $\mathbf{b}_1 = \begin{bmatrix} 1 \\ 0 \\ 0 \end{bmatrix}$, $\mathbf{b}_2 = \begin{bmatrix} 1 \\ 1 \\ 0 \end{bmatrix}$, and $\mathbf{b}_3 = \begin{bmatrix} 1 \\ 2 \\ 2 \end{bmatrix}$. Find the coordinate vector of $\mathbf{u}$ relative to $\widetilde{\mathcal{B}}$.

Solution *To find the coordinate vector of* $\mathbf{u}$ *relative to* $\widetilde{\mathcal{B}}$, *we need to express* $\mathbf{u}$ *as a linear combination of the basis vectors, i.e.,* $\mathbf{u} = x_1\mathbf{b}_1 + x_2\mathbf{b}_2 + x_3\mathbf{b}_3$, *i.e.,*

$$\begin{bmatrix} 5 \\ 2 \\ 0 \end{bmatrix} = x_1 \begin{bmatrix} 1 \\ 0 \\ 0 \end{bmatrix} + x_2 \begin{bmatrix} 1 \\ 1 \\ 0 \end{bmatrix} + x_3 \begin{bmatrix} 1 \\ 2 \\ 2 \end{bmatrix}$$

Solving the system, we obtain the coordinator vector of $\mathbf{u}$ *relative to the basis* $\widetilde{\mathcal{B}}$ *as*

$$[\mathbf{u}]_{\widetilde{\mathcal{B}}} = \begin{bmatrix} x_1 \\ x_2 \\ x_3 \end{bmatrix} = \begin{bmatrix} 3 \\ 2 \\ 0 \end{bmatrix}$$

Remark 9 The coordinator vector of $\mathbf{u}$ relative to the basis $\mathcal{B}$ or $\widetilde{\mathcal{B}}$ of R^3 has 3 entries, which is the same as the dimension of the vector space. For any vector space, its dimension is a contant integer. If a basis of the vector space V has k vectors, then any other bases of V has exactly k vectors.

Theorem 10 *Let V be a k-dimensional subspace of R^n. Any linearly independent set having k vectors of V is a basis for V.*

Dimensions of Column and Null Spaces of a Matrix

The dimension of the column space of a matrix A is called the **rank** of A, and it is denoted by $rank\ A$. We only need to know the pivot columns of A to find a basis for $Col\ A$. The number of pivot columns is the same as the number of vectors in the basis of $Col\ A$. So we know that dim $(Col\ A)$ or $rank\ A$ is the same as the number of pivot columns of A.

Example 11 Determine the rank. $A = \begin{bmatrix} 1 & 2 & 3 & 4 & -2 \\ 0 & 1 & -3 & 0 & 7 \\ 0 & 0 & 0 & 2 & -1 \\ 0 & 0 & 0 & -6 & 3 \end{bmatrix}$.

Solution *We reduce A to its row echelon form*

$$\begin{bmatrix} 1 & 2 & 3 & 4 & -2 \\ 0 & 1 & -3 & 0 & 7 \\ 0 & 0 & 0 & 2 & -1 \\ 0 & 0 & 0 & 0 & 0 \end{bmatrix}$$

which shows the pivot columns are 1, 2, *and* 4. *We can obtain a basis for Col A using the pivot columns of A, i.e.,* $\left\{ \begin{bmatrix} 1 \\ 0 \\ 0 \\ 0 \end{bmatrix}, \begin{bmatrix} 2 \\ 1 \\ 0 \\ 0 \end{bmatrix}, \begin{bmatrix} 4 \\ 0 \\ 2 \\ -6 \end{bmatrix} \right\}$. *The dimension of Col A, i.e., the rank of A, is* 3.

Example 12 Determine the dimension of the null space of A given in the previous example.

Solution *Let's reduce A to its reduced row echelon form*

$$A = \begin{bmatrix} 1 & 2 & 3 & 4 & -2 \\ 0 & 1 & -3 & 0 & 7 \\ 0 & 0 & 0 & 2 & -1 \\ 0 & 0 & 0 & -6 & 3 \end{bmatrix} \sim \begin{bmatrix} 1 & 0 & 9 & 0 & -14 \\ 0 & 1 & -3 & 0 & 7 \\ 0 & 0 & 0 & 1 & -\frac{1}{2} \\ 0 & 0 & 0 & 0 & 0 \end{bmatrix}$$

So the general solution for $A\mathbf{x} = 0$ *is,*

$$\mathbf{x} = x_3 \begin{bmatrix} -9 \\ 3 \\ 1 \\ 0 \\ 0 \end{bmatrix} + x_5 \begin{bmatrix} 14 \\ -7 \\ 0 \\ \frac{1}{2} \\ 1 \end{bmatrix}$$

We can obtain a basis for the null space as $\left\{ \begin{bmatrix} -9 \\ 3 \\ 1 \\ 0 \\ 0 \end{bmatrix}, \begin{bmatrix} 14 \\ -7 \\ 0 \\ \frac{1}{2} \\ 1 \end{bmatrix} \right\}$. *The dimension of the null space of A is* 2.

Theorem 13 *If A is a matrix of n columns, then the sum of* $rank\ A$ *and* $\dim\ (Nul\ A)$ *is equal to n.*

Example 14 Let $A = \begin{bmatrix} 1 & 3 & -1 & 2 \\ 2 & 6 & -2 & 4 \\ 3 & 9 & -3 & 6 \end{bmatrix}$. Decide the $rank\ A$ and $\dim\ (Nul\ A)$.

Solution *We reduce A to echelon form*

$$\begin{bmatrix} 1 & 3 & -1 & 2 \\ 0 & 0 & 0 & 0 \\ 0 & 0 & 0 & 0 \end{bmatrix}$$

which indicates that A has only 1 *pivot column. So* $rank\ A = 1$. *The matrix A has a total of* $n = 4$ *columns, and*

$$rank\ A + \dim\ (Nul\ A) = 4.$$

Hence $\dim\ (Nul\ A) = 3$.

Example 15 Determine the $rank\ A$ if A is a 3×8 matrix and the $Nul\ A$ has 2 free variables.

Solution *If the* $Nul\ A$ *has 2 free variables,* $\dim\ (Nul\ A) = 2$. *The matrix A has a total of* $n = 8$ *columns, and so,*

$$rank\ A + \dim\ (Nul\ A) = 8$$

Therefore, $rank\ A = 6$.

Full Rank and Invertible Matrices

A matrix is **full rank** if its rank is the highest possible value for a matrix of the same size. For any matrix of 3×5, for example, the highest possible value for the rank would be 3, which is the largest possible number of linearly independent columns in any matrix of this size. So when a matrix of 3×5 has the rank 3, it is a full rank matrix. For any $m \times n$ matrix, the highest possible value for the rank would be the smaller of the two values m and n.

Now for an $n \times n$ square matrix A, it is full rank if $rank\ A = n$. A matrix being full rank implies that the matrix is invertible as stated in the following Theorem.

Theorem 16 *Let A be an* $n \times n$ *square matrix. The following statements are equivalent*

(a) $rank\ A = n$
(b) *A has n independent columns*
(c) $Col\ A = R^n$
(d) $\dim\ (Nul\ A) = 0$
(e) $Nul\ A = \{\mathbf{0}\}$
(f) *A is invertible*

Example 17 Let $B = \begin{bmatrix} 1 & -2 & 2 & 3 & -1 \\ 0 & 0 & 5 & 10 & -10 \\ 0 & 0 & 1 & 2 & -2 \end{bmatrix}$. Find a basis for $Col\ B$.

Solution *We apply elementary row operations to reduce the matrix B to its reduced row echelon form*

$$A = \begin{bmatrix} 1 & -2 & 0 & -1 & 3 \\ 0 & 0 & 1 & 2 & -2 \\ 0 & 0 & 0 & 0 & 0 \end{bmatrix}$$

which shows that the pivot columns of B are columns 1 *and* 3*. The row operations will not affect the linear dependence relationship among columns of B. Therefore, a basis for Col B is* $\left\{ \begin{bmatrix} 1 \\ 0 \\ 0 \end{bmatrix}, \begin{bmatrix} 2 \\ 5 \\ 1 \end{bmatrix} \right\}$.

Remark 18 To provide a basis for $Col\ B$ in the above example, we use the pivot columns of B, instead of the columns 1 and 3 of the reduced row echelon form A. The columns of A can't generate $Col\ B$ because all columns of A have 0 in their last entries.

Theorem 19 *The pivot columns of a matrix form a basis for its column space.*

Null Space of a Matrix

The **null space** of a matrix A, denoted by $Nul\ A$, is the set of all solutions of $A\mathbf{x} = \mathbf{0}$. If A is an $m \times n$ matrix, the null space of A is in fact a subspace of R^n.

Example 20 Let $B = \begin{bmatrix} 1 & -2 & 2 & 3 & -1 \\ 0 & 0 & 5 & 10 & -10 \\ 0 & 0 & 1 & 2 & -2 \end{bmatrix}$. Find a basis for $Nul\ B$.

Solution *This matrix B can be row reduced to the matrix A,*

$$A = \begin{bmatrix} 1 & -2 & 0 & -1 & 3 \\ 0 & 0 & 1 & 2 & -2 \\ 0 & 0 & 0 & 0 & 0 \end{bmatrix}$$

The general solution of $A\mathbf{x} = \mathbf{0}$ *has* 3 *free variables, and*

$$\mathbf{x} = \begin{bmatrix} x_1 \\ x_2 \\ x_3 \\ x_4 \\ x_5 \end{bmatrix} = x_2 \begin{bmatrix} 2 \\ 1 \\ 0 \\ 0 \\ 0 \end{bmatrix} + x_4 \begin{bmatrix} 1 \\ 0 \\ -2 \\ 1 \\ 0 \end{bmatrix} + x_5 \begin{bmatrix} -3 \\ 0 \\ 2 \\ 0 \\ 1 \end{bmatrix}$$

So Nul $A =$ *Span* $\left\{ \begin{bmatrix} 2 \\ 1 \\ 0 \\ 0 \\ 0 \end{bmatrix}, \begin{bmatrix} 1 \\ 0 \\ -2 \\ 1 \\ 0 \end{bmatrix}, \begin{bmatrix} -3 \\ 0 \\ 2 \\ 0 \\ 1 \end{bmatrix} \right\}$, *and a basis for Nul A is* $\left\{ \begin{bmatrix} 2 \\ 1 \\ 0 \\ 0 \\ 0 \end{bmatrix}, \begin{bmatrix} 1 \\ 0 \\ -2 \\ 1 \\ 0 \end{bmatrix}, \begin{bmatrix} -3 \\ 0 \\ 2 \\ 0 \\ 1 \end{bmatrix} \right\}$.

Theorem 21 *(i) The null space of an* $m \times n$ *matrix* A *is a subspace of* R^n. *(ii) The column space of an* $m \times n$ *matrix* A *is a subspace of* R^m.

3.5 Exercises

Exercise 22 Decide whether the set $\left\{ \begin{bmatrix} 1 \\ 2 \end{bmatrix}, \begin{bmatrix} 2 \\ 8 \end{bmatrix} \right\}$ give a basis for R^2.

Exercise 23 Decide whether the set $\left\{ \begin{bmatrix} 1 \\ 0 \\ 1 \end{bmatrix}, \begin{bmatrix} 0 \\ 1 \\ 1 \end{bmatrix} \right\}$ give a basis for R^3.

Exercise 24 Decide whether the set $\left\{ \begin{bmatrix} 1 \\ 0 \\ 1 \end{bmatrix}, \begin{bmatrix} 0 \\ 1 \\ 1 \end{bmatrix}, \begin{bmatrix} 1 \\ 1 \\ 0 \end{bmatrix} \right\}$ give a basis for R^3.

Exercise 25 Decide whether the set $\left\{ \begin{bmatrix} 1 \\ 0 \\ 1 \end{bmatrix}, \begin{bmatrix} 0 \\ 1 \\ 1 \end{bmatrix}, \begin{bmatrix} 1 \\ 1 \\ 0 \end{bmatrix}, \begin{bmatrix} 2 \\ 5 \\ 5 \end{bmatrix} \right\}$ give a basis for R^3.

Exercise 26 Decide whether the set $\left\{ \begin{bmatrix} 1 \\ 0 \\ 1 \\ 1 \end{bmatrix}, \begin{bmatrix} 0 \\ 1 \\ 2 \\ 3 \end{bmatrix}, \begin{bmatrix} 1 \\ 1 \\ 2 \\ 2 \end{bmatrix}, \begin{bmatrix} 4 \\ -4 \\ 1 \\ 2 \end{bmatrix} \right\}$ give a basis for R^4.

Exercise 27 Use the basis $\mathcal{B} = \left\{ \begin{bmatrix} 1 \\ 1 \end{bmatrix}, \begin{bmatrix} 1 \\ 3 \end{bmatrix} \right\}$ for R^2. Find the coordinate vector of $\mathbf{u} = \begin{bmatrix} 2 \\ 8 \end{bmatrix}$ relative to $\mathcal{B}$.

Exercise 28 Use the basis $\mathcal{B} = \left\{ \begin{bmatrix} 1 \\ 2 \end{bmatrix}, \begin{bmatrix} 2 \\ 3 \end{bmatrix} \right\}$ for R^2. Find the coordinate vector of $\mathbf{u} = \begin{bmatrix} 2 \\ 8 \end{bmatrix}$ relative to $\mathcal{B}$.

Exercise 29 Use the basis $\mathcal{B} = \left\{ \begin{bmatrix} 2 \\ 0 \\ 0 \end{bmatrix}, \begin{bmatrix} 2 \\ -1 \\ 0 \end{bmatrix}, \begin{bmatrix} 2 \\ -1 \\ 1 \end{bmatrix} \right\}$ for R^3. Find the coordinate vector of $\mathbf{u} = \begin{bmatrix} 2 \\ 8 \\ 5 \end{bmatrix}$ relative to $\mathcal{B}$.

Exercise 30 Use the basis $\mathcal{B} = \left\{ \begin{bmatrix} 1 \\ 0 \\ 0 \end{bmatrix}, \begin{bmatrix} 1 \\ 2 \\ 0 \end{bmatrix}, \begin{bmatrix} 1 \\ 2 \\ 3 \end{bmatrix} \right\}$ for R^3. Find the coordinate vector of $\mathbf{u} = \begin{bmatrix} 2 \\ 8 \\ 5 \end{bmatrix}$ relative to $\mathcal{B}$.

Exercise 31 Given $A = \begin{bmatrix} 1 & 2 \\ 0 & 1 \\ 1 & 1 \end{bmatrix}$. (i) Find a basis for $Col\ A$. (ii) Find a basis for $Nul\ A$.

Exercise 32 Given $A = \begin{bmatrix} 1 & 0 & -1 & 1 \\ 0 & 1 & 2 & -2 \\ 0 & 0 & 0 & 0 \end{bmatrix}$. (i) Find a basis for $Col\ A$. (ii) Find a basis for $Nul\ A$.

Exercise 33 Determine the rank of $A = \begin{bmatrix} 1 & 0 & 2 \\ 0 & 1 & 0 \\ 0 & 1 & 1 \end{bmatrix}$. Is A a full rank matrix?

Exercise 34 Determind the rank of $A = \begin{bmatrix} 1 & 0 & -3 & 1 \\ 0 & 2 & 2 & -2 \\ 0 & 0 & 0 & 8 \end{bmatrix}$. Is A a full rank matrix?

Determinants 4

4.1 Introduction to Determinants

The determinant has been introduced for 2×2 matrices. Recall that the determinant of

$$A_{2\times 2} = \begin{bmatrix} a_{11} & a_{12} \\ a_{21} & a_{22} \end{bmatrix}$$

is the number

$$\det\ A_{2\times 2} = a_{11}a_{22} - a_{12}a_{21}$$

The determinant of a matrix is also denoted by vertical lines, for example,

$$\begin{vmatrix} a_{11} & a_{12} \\ a_{21} & a_{22} \end{vmatrix}$$

means the same as $\det \begin{bmatrix} a_{11} & a_{12} \\ a_{21} & a_{22} \end{bmatrix}$. We define the determinant of $A_{1\times 1} = [a_{11}]$, a 1×1 matrix, to be $\det\ [a_{11}] = a_{11}$. Then the determinant formula for a 2×2 matrix $A_{2\times 2} = \begin{bmatrix} a_{11} & a_{12} \\ a_{21} & a_{22} \end{bmatrix}$ can be written as,

$$\det\ A_{2\times 2} = a_{11} \det\ [a_{22}] - a_{12} \det\ [a_{21}]$$

which relates the determinant of a 2×2 matrix to determinants of 1×1 matrices. Extending this recursive pattern to a 3×3 matrix,

$$A_{3\times 3} = \begin{bmatrix} a_{11} & a_{12} & a_{13} \\ a_{21} & a_{22} & a_{23} \\ a_{31} & a_{32} & a_{33} \end{bmatrix}$$

H. Tian, *Linear Algebra*, Synthesis Lectures on Mathematics & Statistics,
https://doi.org/10.1007/978-3-031-84647-2_4

we define

$$\det\ A_{3\times3} = a_{11} \det A_{11} - a_{12} \det A_{12} + a_{13} \det A_{13}$$

where A_{11}, A_{12}, and A_{13} are smaller size matrices corresponding to the entries a_{11}, a_{12}, and a_{13} respectively, as given below,

$$A_{11} = \begin{bmatrix} a_{22} & a_{23} \\ a_{32} & a_{33} \end{bmatrix},\ A_{12} = \begin{bmatrix} a_{21} & a_{23} \\ a_{31} & a_{33} \end{bmatrix},\ A_{13} = \begin{bmatrix} a_{21} & a_{22} \\ a_{31} & a_{32} \end{bmatrix}.$$

Example 1 Calculate the determinant of $\begin{bmatrix} 3 & 0 & -1 \\ 0 & 1 & 2 \\ 1 & 2 & 1 \end{bmatrix}$.

Solution *Let's apply the determinant formula to the* 3×3 *matrix,*

$$\begin{aligned} \det\ A_{3\times3} &= a_{11} \det A_{11} - a_{12} \det A_{12} + a_{13} \det A_{13} \\ &= 3 \det \begin{bmatrix} 1 & 2 \\ 2 & 1 \end{bmatrix} - 0 \det \begin{bmatrix} 0 & 2 \\ 1 & 1 \end{bmatrix} + (-1) \det \begin{bmatrix} 0 & 1 \\ 1 & 2 \end{bmatrix} \\ &= 3\,(-3) - 0\,(-2) + (-1)\,(-1) \\ &= -8 \end{aligned}$$

For a general $n \times n$ matrix with $n \geq 2$,

$$A_{n\times n} = \begin{bmatrix} a_{11} & a_{12} & \cdots & a_{1n} \\ a_{21} & a_{22} & \cdots & a_{2n} \\ \vdots & \vdots & \vdots & \vdots \\ a_{n1} & a_{n2} & \cdots & a_{nn} \end{bmatrix}$$

Its **determinant** is defined as

$$\det\ A_{n\times n} = a_{11} \det A_{11} - a_{12} \det A_{12} + \cdots + (-1)^{1+n} a_{1n} \det A_{1n} = \sum_{j=1}^{n} (-1)^{1+j} a_{1j} \det A_{1j}$$

where A_{11}, A_{12}, $\ldots$, A_{1n} are smaller size $(n-1) \times (n-1)$ matrices corresponding to the entries a_{11}, a_{12}, $\ldots$, a_{1n} respectively, as given below,

$$A_{11} = \begin{bmatrix} a_{22} & \cdots & a_{2n} \\ \vdots & \vdots & \vdots \\ a_{n2} & \cdots & a_{nn} \end{bmatrix},\ \ldots,\ A_{1n} = \begin{bmatrix} a_{21} & \cdots & a_{2(n-1)} \\ \vdots & \vdots & \vdots \\ a_{n1} & \cdots & a_{n(n-1)} \end{bmatrix}.$$

The A_{1j} for any $j = 1, 2, \ldots, n$, is the $(n-1) \times (n-1)$ matrix that is obtained from removing the 1st row and jth column of $A_{n\times n}$.

We can slightly rewrite the above formula and give a different representation of the determinant,

$$\det A_{n\times n} = a_{11}C_{11} + a_{12}C_{12} + \cdots + a_{1n}C_{1n} = \sum_{j=1}^{n} a_{1j}C_{1j}$$

where

$$C_{11} = \det A_{11}, \ C_{12} = -\det A_{12}, \ \ldots, \ C_{1n} = (-1)^{1+n}\det A_{1n}.$$

The C_{1j} is the $(1, j)$-cofactor of A, with $j = 1, 2, \ldots, n$. This alternative formula for the determinant is called a **cofactor expansion** across the first row of A. The alternating signs in the first formula are absorbed in the cofactors.

In fact, a theorem shows that the determinant of an $n \times n$ matrix A can be calculated using any row or column.

Theorem 2 *The determinant of an $n \times n$ matrix A can be calculated using cofactor expansion across any row or down any column of A. More specifically, the cofactor expansion across the ith row of A is,*

$$\det A = a_{i1}C_{i1} + a_{i2}C_{i2} + \cdots + a_{in}C_{in} = \sum_{j=1}^{n} a_{ij}C_{ij}$$

where

$$C_{i1} = (-1)^{i+1}\det A_{i1}, \ C_{i2} = (-1)^{i+2}\det A_{i2}, \ \ldots, \ C_{in} = (-1)^{i+n}\det A_{in}.$$

and the cofactor expansion down the jth column of A is,

$$\det A = a_{1j}C_{1j} + a_{2j}C_{2j} + \cdots + a_{nj}C_{nj} = \sum_{i=1}^{n} a_{ij}C_{ij}$$

where

$$C_{1j} = (-1)^{1+j}\det A_{1j}, \ C_{2j} = (-1)^{2+j}\det A_{2j}, \ \ldots, \ C_{nj} = (-1)^{n+j}\det A_{nj}.$$

In summary, the cofactors

$$C_{ij} = (-1)^{i+j}\det A_{ij}$$

with the part $(-1)^{i+j}$ indicating the checkerboard pattern of signs in front of $\det A_{ij}$.

Corollary 3 *The determinant of an $n \times n$ matrix A is 0 if A has a row of zeros or a column of zeros.*

Example 4 Use the cofactor expansion to calculate the determinant of the matrix $A = \begin{bmatrix} 2 & 1 & -1 \\ 1 & 0 & 3 \\ 1 & 6 & 0 \end{bmatrix}$.

Solution *Let's use the cofactor expansion down the third column of A,*

$$\begin{aligned}
\det A &= a_{13}C_{13} + a_{23}C_{23} + a_{33}C_{33} \\
&= a_{13}(-1)^{1+3}\det A_{13} + a_{23}\,(-1)^{2+3}\det A_{23} + a_{33}\,(-1)^{3+3}\det A_{33} \\
&= a_{13}\begin{vmatrix} 1 & 0 \\ 1 & 6 \end{vmatrix} - a_{23}\begin{vmatrix} 2 & 1 \\ 1 & 6 \end{vmatrix} + a_{33}\begin{vmatrix} 2 & 1 \\ 1 & 0 \end{vmatrix} \\
&= (-1)\begin{vmatrix} 1 & 0 \\ 1 & 6 \end{vmatrix} - (3)\begin{vmatrix} 2 & 1 \\ 1 & 6 \end{vmatrix} + (0)\begin{vmatrix} 2 & 1 \\ 1 & 0 \end{vmatrix} \\
&= -6 - 33 + 0 \\
&= -39
\end{aligned}$$

Example 5 Calculate the determinant of

$$A = \begin{bmatrix} 5 & -7 & 2 & 2 \\ 0 & 3 & 0 & -4 \\ -5 & -8 & 0 & 3 \\ 0 & 5 & 0 & -6 \end{bmatrix}.$$

Solution *Let's use the cofactor expansion down the* 3*rd column of A, due to many* 0 *entries in this column.*

$$\begin{aligned}
\det A &= a_{13}C_{13} + a_{23}C_{23} + a_{33}C_{33} + a_{43}C_{43} \\
&= 2C_{13} + 0C_{23} + 0C_{33} + 0C_{43} \\
&= 2\,(-1)^{1+3}\det A_{13} \\
&= 2\begin{vmatrix} 0 & 3 & -4 \\ -5 & -8 & 3 \\ 0 & 5 & -6 \end{vmatrix}
\end{aligned}$$

Now, the 1st column is a good choice for the cofactor expansion due to more 0 entries in the column.

$$\det A = 2\begin{vmatrix} 0 & 3 & -4 \\ -5 & -8 & 3 \\ 0 & 5 & -6 \end{vmatrix}$$

$$= 2\,(-5)\,(-1)^{2+1}\begin{vmatrix} 3 & -4 \\ 5 & -6 \end{vmatrix}$$

$$= 2\,(-5)\,(-1)\,(2)$$

$$= 20$$

Example 6 Calculate the determinant of the triangular matrices,

$$A = \begin{bmatrix} 3 & 0 & 0 & 0 \\ 1 & 2 & 0 & 0 \\ 9 & -1 & 7 & 0 \\ 2 & 6 & 8 & 4 \end{bmatrix}, \qquad A^T = \begin{bmatrix} 3 & 1 & 9 & 2 \\ 0 & 2 & -1 & 6 \\ 0 & 0 & 7 & 8 \\ 0 & 0 & 0 & 4 \end{bmatrix}.$$

Solution *For the matrix A, we start with the cofactor expansion across the 1st row and then continue the similar fashion until we finish.*

$$\det A = \begin{vmatrix} 3 & 0 & 0 & 0 \\ 1 & 2 & 0 & 0 \\ 9 & -1 & 7 & 0 \\ 2 & 6 & 8 & 4 \end{vmatrix}$$

$$= 3C_{11} + 0C_{12} + 0C_{13} + 0C_{14}$$

$$= 3\begin{vmatrix} 2 & 0 & 0 \\ -1 & 7 & 0 \\ 6 & 8 & 4 \end{vmatrix}$$

$$= 3\,(2)\begin{vmatrix} 7 & 0 \\ 8 & 4 \end{vmatrix}$$

$$= 3\,(2)\,(7)\,(4)$$

$$= 168$$

For the matrix A^T, we start with the cofactor expansion across the 1st column and then continue the similar fashion until we finish.

$$\det A^T = \begin{vmatrix} 3 & 1 & 9 & 2 \\ 0 & 2 & -1 & 6 \\ 0 & 0 & 7 & 8 \\ 0 & 0 & 0 & 4 \end{vmatrix}$$

$$= 3C_{11} + 0C_{21} + 0C_{31} + 0C_{41}$$

$$= 3\begin{vmatrix} 2 & -1 & 6 \\ 0 & 7 & 8 \\ 0 & 0 & 4 \end{vmatrix}$$

$$= 3\,(2)\begin{vmatrix} 7 & 8 \\ 0 & 4 \end{vmatrix}$$
$$= 3\,(2)\,(7)\,(4)$$
$$= 168$$

Now that we know determinants of matrices can be calculated using cofactor expansion across any row and down any column. Let's look at a square $n \times n$ matrix A in its column format, i.e., A is represented with its n columns,

$$A = \begin{bmatrix} \mathbf{a}_1 & \cdots & \mathbf{a}_{j-1} & \mathbf{a}_j & \mathbf{a}_{j+1} & \cdots & \mathbf{a}_n \end{bmatrix}$$

The determinant of A can be calculated using the cofactor expansion down the jth column. But if we allow the jth column to be a vector variable $\mathbf{x}$, the determinant will be a function of $\mathbf{x}$, which we denote it by $D(\mathbf{x})$.

$$D(\mathbf{x}) = \det\begin{bmatrix} \mathbf{a}_1 & \cdots & \mathbf{a}_{j-1} & \mathbf{x} & \mathbf{a}_{j+1} & \cdots & \mathbf{a}_n \end{bmatrix}$$

This determinant function satisfies the linear properties:

$$\begin{aligned} &\text{(i) } D(c\mathbf{x}) = cD(\mathbf{x}) \\ &\text{(ii) } D(\mathbf{x}+\mathbf{y}) = D(\mathbf{x}) + D(\mathbf{y}) \end{aligned}$$

which is the same as,

$$\begin{aligned} &\text{(i) } \det\begin{bmatrix} \mathbf{a}_1 & \cdots & \mathbf{a}_{j-1} & c\mathbf{x} & \mathbf{a}_{j+1} & \cdots & \mathbf{a}_n \end{bmatrix} = c\det\begin{bmatrix} \mathbf{a}_1 & \cdots & \mathbf{a}_{j-1} & \mathbf{x} & \mathbf{a}_{j+1} & \cdots & \mathbf{a}_n \end{bmatrix} \\ &\text{(ii) } \det\begin{bmatrix} \mathbf{a}_1 & \cdots & \mathbf{a}_{j-1} & \mathbf{x}+\mathbf{y} & \mathbf{a}_{j+1} & \cdots & \mathbf{a}_n \end{bmatrix} = \det\begin{bmatrix} \mathbf{a}_1 & \cdots & \mathbf{a}_{j-1} & \mathbf{x} & \mathbf{a}_{j+1} & \cdots & \mathbf{a}_n \end{bmatrix} \\ &\qquad\qquad + \det\begin{bmatrix} \mathbf{a}_1 & \cdots & \mathbf{a}_{j-1} & \mathbf{y} & \mathbf{a}_{j+1} & \cdots & \mathbf{a}_n \end{bmatrix} \end{aligned}$$

Example 7 According to the linearity property of the determinant function, the following two equations are true.

$$\begin{vmatrix} 1 & 3c \\ 2 & 4c \end{vmatrix} = c\begin{vmatrix} 1 & 3 \\ 2 & 4 \end{vmatrix},$$

$$\begin{vmatrix} 1 & x_1 + y_1 \\ 2 & x_2 + y_2 \end{vmatrix} = \begin{vmatrix} 1 & x_1 \\ 2 & x_2 \end{vmatrix} + \begin{vmatrix} 1 & y_1 \\ 2 & y_2 \end{vmatrix}.$$

Next, we will calculate determinants of triangular matrices. They are easy to work with because of its special structure and many 0 entries. The following Theorem can be proved true for any triangular matrices.

Theorem 8 *The determinant of a triangular matrix is the product of its main diagonal entries.*

Corollary 9 *The determinant of a triangular matrix is* 0 *if there exists a* 0 *entry on the main diagonal of the matrix.*

Example 10 Calculate the determinant of the matrix $B = \begin{bmatrix} 1 & -2 & 2 \\ 0 & 0 & 5 \\ 0 & 0 & 9 \end{bmatrix}$.

Solution *This is a triangular matrix and it has a* 0 *entry on its main diagonal. So* $\det B = 0$.

4.1 Exercises

Exercise 11 Calculate the determinant of each matrix.

$$A = \begin{bmatrix} 1 & 0 & 2 \\ 0 & 1 & 0 \\ 0 & 1 & 5 \end{bmatrix}, \quad B = \begin{bmatrix} 1 & 0 & 0 \\ 0 & 1 & 1 \\ 2 & 0 & 5 \end{bmatrix}$$

Exercise 12 Calculate the determinant of each matrix

$$A = \begin{bmatrix} 2 & 5 & 3 \\ 3 & 2 & 1 \\ 1 & 0 & -1 \end{bmatrix}, \quad B = \begin{bmatrix} 1 & 0 & -1 \\ 3 & 2 & 1 \\ 2 & 5 & 3 \end{bmatrix}, \quad C = \begin{bmatrix} 2 & 0 & -2 \\ 3 & 2 & 1 \\ 2 & 5 & 3 \end{bmatrix}$$

Exercise 13 Calculate the determinant of each matrix.

$$A = \begin{bmatrix} 1 & 6 & 2 & 8 \\ 1 & 0 & -4 & 0 \\ 0 & 0 & 5 & 0 \\ 2 & 5 & 6 & 2 \end{bmatrix}, \quad B = \begin{bmatrix} 0 & 0 & 5 & 0 \\ 1 & 0 & -4 & 0 \\ 1 & 6 & 2 & 8 \\ 2 & 5 & 6 & 2 \end{bmatrix}, \quad C = \begin{bmatrix} 0 & 0 & 5 & 0 \\ 2 & 0 & -8 & 0 \\ 1 & 6 & 2 & 8 \\ 2 & 5 & 6 & 2 \end{bmatrix}$$

Exercise 14 Calculate the determinant of each triangular matrix.

$$A = \begin{bmatrix} 5 & 0 & 0 & 0 \\ 1 & -1 & 0 & 0 \\ 0 & -1 & 4 & 0 \\ 6 & 0 & 3 & 8 \end{bmatrix}, \quad A^T = \begin{bmatrix} 5 & 1 & 0 & 6 \\ 0 & -1 & -1 & 0 \\ 0 & 0 & 4 & 3 \\ 0 & 0 & 0 & 8 \end{bmatrix}.$$

Exercise 15 Calculate the determinant of each triangular matrix.

$$A = \begin{bmatrix} 1 & 2 & 3 & 4 & -2 \\ 0 & 1 & -3 & 0 & 7 \\ 0 & 0 & 0 & 2 & -1 \\ 0 & 0 & 0 & -6 & 3 \\ 0 & 0 & 0 & 0 & 4 \end{bmatrix}, \qquad A^T = \begin{bmatrix} 1 & 0 & 0 & 0 & 0 \\ 2 & 1 & 0 & 0 & 0 \\ 3 & -3 & 0 & 0 & 0 \\ 4 & 0 & 2 & -6 & 0 \\ -2 & 7 & -1 & 3 & 4 \end{bmatrix}$$

4.2 Properties of Determinants

Row Operations

When a square matrix is in echelon form, its determinant is easy for us to determine. We know we can always reduce a matrix A to its echelon form B using a sequence of elementary row operations. However, the determinant of B can be different than the determinant of A.

Let's consider a square matrix A and $\widetilde{A}$, with $\widetilde{A}$ being the matrix we obtain after applying an elementary row operation to A. Will the determinants of A and $\widetilde{A}$ be the same?

We know there are 3 different ways to get $\widetilde{A}$ from A : (i) interchange any two rows of A; (ii) multiply a row of A by a scalar; (iii) add a multiple of a row of A to another row. In the following, we use a small 2×2 matrix

$$A = \begin{bmatrix} a_{11} & a_{12} \\ a_{21} & a_{22} \end{bmatrix}$$

and apply one of the three row operations to get $\widetilde{A}$. Let's compare the determinants of A and $\widetilde{A}$. The A has the determinant

$$\det A = a_{11}a_{22} - a_{12}a_{21}$$

(i) Interchanging two rows of A to obtain $\widetilde{A}$,

$$\widetilde{A} = \begin{bmatrix} a_{21} & a_{22} \\ a_{11} & a_{12} \end{bmatrix}$$

We have

$$\det \widetilde{A} = a_{12}a_{21} - a_{11}a_{22}$$

which means

$$\det \widetilde{A} = -\det A$$

(ii) Multiplying a row of A by a scalar d to obtain $\widetilde{A}$

$$\widetilde{A} = \begin{bmatrix} da_{11} & da_{12} \\ a_{21} & a_{22} \end{bmatrix} \text{ or } \widetilde{A} = \begin{bmatrix} a_{11} & a_{12} \\ da_{21} & da_{22} \end{bmatrix}$$

We have

$$\det \widetilde{A} = da_{11}a_{22} - da_{12}a_{21}$$

which means

$$\det \widetilde{A} = d \det A$$

(iii) Adding a multiple of a row of A to another row to obtain $\widetilde{A}$ (adding a multiple of 1st row of A to the 2nd row, or adding a multiple of 2nd row of A to the 1st row)

$$\widetilde{A} = \begin{bmatrix} a_{11} & a_{12} \\ ca_{11} + a_{21} & ca_{12} + a_{22} \end{bmatrix} \text{ or } \widetilde{A} = \begin{bmatrix} ca_{21} + a_{11} & ca_{22} + a_{12} \\ a_{21} & a_{22} \end{bmatrix}$$

We have

$$\det \widetilde{A} = a_{11}a_{22} - a_{12}a_{21}$$

which means

$$\det \widetilde{A} = \det A$$

The generalized version of the above results can be proved and are given in the following theorem.

Theorem 1 *Let A be an $n \times n$ square matrix and $\widetilde{A}$ be the matrix obtained by applying an elementary row operation to A.*

(i) If $\widetilde{A}$ is obtained by interchanging two rows of A, then $\det \widetilde{A} = -\det A$.
(ii) If $\widetilde{A}$ is obtained by multiplying a row of A by a scalar d, then $\det \widetilde{A} = d \det A$.
(iii) If $\widetilde{A}$ is obtained by adding a multiple of a row of A to another row, then $\det \widetilde{A} = \det A$.

Applying the above Theorem, we can calculate determinants of matrices using the echelon form approach.

Example 2 Calculate the determinant of

$$A = \begin{bmatrix} 2 & 1 & 0 \\ 1 & 2 & 6 \\ 1 & 0 & 1 \end{bmatrix}.$$

Solution *We first interchange the first and the third rows of A, and then do the replacements of the second and third rows.*

$$\det A = \begin{vmatrix} 2 & 1 & 0 \\ 1 & 2 & 6 \\ 1 & 0 & 1 \end{vmatrix} = - \begin{vmatrix} 1 & 0 & 1 \\ 1 & 2 & 6 \\ 2 & 1 & 0 \end{vmatrix} = - \begin{vmatrix} 1 & 0 & 1 \\ 0 & 2 & 5 \\ 0 & 1 & -2 \end{vmatrix}$$

Next, we interchange the second and the third rows (which results in another negative sign), and then do the replacement of the third row.

$$\det A = \begin{vmatrix} 1 & 0 & 1 \\ 0 & 1 & -2 \\ 0 & 2 & 5 \end{vmatrix} = \begin{vmatrix} 1 & 0 & 1 \\ 0 & 1 & -2 \\ 0 & 0 & 9 \end{vmatrix} = 9.$$

Example 3 Calculate the determinant of

$$C = \begin{bmatrix} 2 & 1 & 0 \\ 3 & 6 & 18 \\ 1 & 0 & 1 \end{bmatrix}.$$

Solution *We first multiply the second row of C by* $1/3$, *and then we can proceed as in the previous example.*

$$\det C = \begin{vmatrix} 2 & 1 & 0 \\ 3 & 6 & 18 \\ 1 & 0 & 1 \end{vmatrix} = 3 \begin{vmatrix} 2 & 1 & 0 \\ 1 & 2 & 6 \\ 1 & 0 & 1 \end{vmatrix} = 3\,(9) = 27.$$

Example 4 Calculate the determinant of

$$A = \begin{bmatrix} 1 & 2 & 3 \\ -3 & -6 & -9 \\ 2 & 0 & 2 \end{bmatrix}.$$

Solution *By doing the replacement of adding* 3 *times the first row to the second row of* A, *we obtain*

$$\det A = \begin{vmatrix} 1 & 2 & 3 \\ -3 & -6 & -9 \\ 2 & 0 & 2 \end{vmatrix} = \begin{vmatrix} 1 & 2 & 3 \\ 0 & 0 & 0 \\ 2 & 0 & 2 \end{vmatrix}$$

So $\det A = 0$ *by the cofactor expansion across the second row of zeros.*

Remark 5 The methods of row reduction and cofactor expansion can be combined to find determinants of matrices.

Remark 6 When a matrix has a row that is a multiple of another row, the matrix has a determinant of 0.

We recall that an $n \times n$ matrix is invertible if and only if it can be row reduced to an identity matrix, i.e., if and only if it has n pivots. Then we have the following equivalent statements for an $n \times n$ matrix A.

(a) A is invertible
(b) A can be row reduced to a matrix having n pivots.
(c) $\det A \neq 0$

Example 7 Determine if $A = \begin{bmatrix} 5 & 8 & 8 \\ 0 & -5 & 6 \\ 0 & -5 & 6 \end{bmatrix}$ is invertible.

Solution *The second and third rows of A are identical. The matrix can be row reduced to*

$$B = \begin{bmatrix} 5 & 8 & 8 \\ 0 & -5 & 6 \\ 0 & 0 & 0 \end{bmatrix}$$

So $\det A = 0$, *and A is not invertible.*

A few other important properties of determinants are listed below.

Theorem 8 *For a square matrix A,* $\det A^T = \det A$.

Theorem 9 *For two* $n \times n$ *square matrices A and B,* $\det AB = \det A \det B$

Corollary 10 *If A is invertible, then* $\det\left(A^{-1}\right) = \dfrac{1}{\det A}$.

Corollary 11 *If a square matrix A satisfying* $A^2 = I$. *Then* $\det A = \pm 1$.

Example 12 Let $A = \begin{bmatrix} 1 & -1 \\ 0 & 2 \end{bmatrix}$ and $B = \begin{bmatrix} 2 & 0 \\ 1 & 1 \end{bmatrix}$. Then we have $AB = \begin{bmatrix} 1 & -1 \\ 2 & 2 \end{bmatrix}$. According to $\det AB = \det A \det B$, we know that

$$\det AB = \begin{vmatrix} 1 & -1 \\ 2 & 2 \end{vmatrix} = \det A \det B = \begin{vmatrix} 1 & -1 \\ 0 & 2 \end{vmatrix} \begin{vmatrix} 2 & 0 \\ 1 & 1 \end{vmatrix} = 2\,(2) = 4.$$

4.2 Exercises

Exercise 13 Calculate the determinant of each matrix using the row reduction method.

$$A = \begin{bmatrix} 1 & 0 & 2 \\ 0 & 1 & 0 \\ 0 & 1 & 5 \end{bmatrix}, \quad B = \begin{bmatrix} 1 & 0 & 0 \\ 0 & 1 & 1 \\ 2 & 0 & 5 \end{bmatrix}$$

Exercise 14 Calculate the determinant of each matrix using the combination of the methods of row reduction and cofactor expansion.

$$A = \begin{bmatrix} 2 & 5 & 3 \\ 3 & 2 & 1 \\ 1 & 0 & -1 \end{bmatrix}, \quad B = \begin{bmatrix} 1 & 0 & -1 \\ 3 & 2 & 1 \\ 2 & 5 & 3 \end{bmatrix}, \quad C = \begin{bmatrix} 2 & 0 & -2 \\ 3 & 2 & 1 \\ 2 & 5 & 3 \end{bmatrix}$$

Exercise 15 Use the row reduction method to find the determinant of each matrix, and decide if each matrix is invertible.

$$A = \begin{bmatrix} 1 & 2 & 3 \\ 0 & 5 & -4 \\ 3 & 7 & 4 \end{bmatrix}, \quad B = \begin{bmatrix} 0 & 2 & 1 & 1 \\ -2 & 2 & -2 & -3 \\ 1 & -1 & 0 & 1 \\ 0 & 0 & 9 & 3 \end{bmatrix}, \quad C = \begin{bmatrix} 1 & -2 & 5 & 2 \\ 0 & 0 & 3 & 0 \\ 2 & -6 & -7 & 5 \\ 5 & 0 & 4 & 4 \end{bmatrix}$$

Exercise 16 Calculate the determinant of each matrix

$$A = \begin{bmatrix} 2 & 10 & 0 & 1 \\ 0 & 0 & 5 & 5 \\ 2 & 8 & 2 & -6 \\ 0 & 0 & -2 & 1 \end{bmatrix}, \quad B = \begin{bmatrix} 2 & -1 & 0 & 0 \\ -1 & 2 & -1 & 0 \\ 0 & -1 & 2 & -1 \\ 0 & 0 & -1 & 2 \end{bmatrix}$$

4.3 Applications of Determinants

Cramer's Rule

When a square $n \times n$ matrix A is invertible, the system $A\mathbf{x} = \mathbf{b}$ has one and only one solution for any $\mathbf{b}$ in R^n. Cramer's rule allows us to represent the ith entry x_i, $i = 1, 2, \ldots, n$ of $\mathbf{x}$.

Theorem 1 (Cramer's Rule) *Let A be an invertible $n \times n$ matrix, and let $\mathbf{b} \in R^n$. Then the system $A\mathbf{x} = \mathbf{b}$ has a unique solution $\mathbf{x}$, for which the ith entry x_i, $i = 1, 2, \ldots, n$ is given by*

$$x_i = \frac{\det A_i\,(\mathbf{b})}{\det A}$$

where $A_i\,(\mathbf{b})$ is the matrix obtained by replacing the ith column of A with the vector $\mathbf{b}$.

Proof Note that the vector equation form of the matrix equation $A\mathbf{x} = \mathbf{b}$, i.e., $[\mathbf{a}_1\ \mathbf{a}_2\ \cdots\ \mathbf{a}_{i-1}\ \mathbf{a}_i\ \mathbf{a}_{i+1}\ \cdots\ \mathbf{a}_n]\,\mathbf{x} = \mathbf{b}$ is,

$$x_1\mathbf{a}_1 + \cdots x_{i-1}\mathbf{a}_{i-1} + x_i\mathbf{a}_i + x_{i+1}\mathbf{a}_{i+1} + \cdots + x_n\mathbf{a}_n = \mathbf{b}$$

We know that $A_i\,(\mathbf{b})$ is the matrix obtained by replacing the ith column of A with the vector $\mathbf{b}$, so using the linearity property of the determinant function, we have

$$\begin{aligned}\det A_i(\mathbf{b}) &= \det\begin{bmatrix}\mathbf{a}_1 & \mathbf{a}_2 & \cdots & \mathbf{a}_{i-1} & \mathbf{b} & \mathbf{a}_{i+1} & \cdots & \mathbf{a}_n\end{bmatrix} \\ &= \det\begin{bmatrix}\mathbf{a}_1 & \mathbf{a}_2 & \cdots & \mathbf{a}_{i-1} & x_i\mathbf{a_i} & \mathbf{a}_{i+1} & \cdots & \mathbf{a}_n\end{bmatrix} \\ &= x_i \det\begin{bmatrix}\mathbf{a}_1 & \mathbf{a}_2 & \cdots & \mathbf{a}_{i-1} & \mathbf{a_i} & \mathbf{a}_{i+1} & \cdots & \mathbf{a}_n\end{bmatrix} \\ &= x_i \det A\end{aligned}$$

Therefore,

$$x_i = \frac{\det A_i(\mathbf{b})}{\det A}, \text{ for } i = 1, 2, \ldots, n$$

■

Example 2 Let's solve $A\mathbf{x} = \mathbf{b}$, specifically,

$$\begin{bmatrix}-5 & 3 \\ 3 & -1\end{bmatrix}\begin{bmatrix}x_1 \\ x_2\end{bmatrix} = \begin{bmatrix}9 \\ -5\end{bmatrix}$$

Thus,

$$\det A = \begin{vmatrix}-5 & 3 \\ 3 & -1\end{vmatrix} = -4$$

$$\det A_1\mathbf{b} = \begin{vmatrix}9 & 3 \\ -5 & -1\end{vmatrix} = 6$$

$$\det A_2\mathbf{b} = \begin{vmatrix}-5 & 9 \\ 3 & -5\end{vmatrix} = -2$$

Therefore, using Cramer's Rule, we obtain

$$\begin{aligned}x_1 &= \frac{\det A_1(\mathbf{b})}{\det A} = \frac{6}{-4} = -\frac{3}{2} \\ x_2 &= \frac{\det A_2(\mathbf{b})}{\det A} = \frac{-2}{-4} = \frac{1}{2}.\end{aligned}$$

Inverse Formula

Next, we use Cramer's rule to derive a formula for the inverse matrix of A when $\det A \neq 0$.

Theorem 3 *Let A be an $n \times n$ invertible matrix and let $C = \begin{bmatrix}C_{ij}\end{bmatrix}$ be the cofactor matrix of A. Then*

$$A^{-1} = \frac{1}{\det A}C^T$$

Proof If $H = \begin{bmatrix}\mathbf{h}_1 & \mathbf{h}_2 & \cdots & \mathbf{h}_{j-1} & \mathbf{h}_j & \mathbf{h}_{j+1} & \cdots & \mathbf{h}_n\end{bmatrix}$ is the inverse matrix of A, then $AH = I$, which means,

$$A\left[\mathbf{h}_1\ \mathbf{h}_2\ \cdots\ \mathbf{h}_{j-1}\ \mathbf{h}_j\ \mathbf{h}_{j+1}\ \cdots\ \mathbf{h}_n\right] = \left[\mathbf{e}_1\ \mathbf{e}_2\ \cdots\ \mathbf{e}_{j-1}\ \mathbf{e}_j\ \mathbf{e}_{j+1}\ \cdots\ \mathbf{e}_n\right]$$

Thus

$$A\mathbf{h}_j = \mathbf{e}_j, \text{ for } j = 1,\ 2,\ \ldots,\ n$$

The $\mathbf{h}_j$, the jth column of A^{-1}, can be found by Cramer's rule. The ith entry of $\mathbf{h}_j$, which is the entry of A^{-1} at the ith row and jth column, is calculated as follows using Cramer's rule,

$$\left(A^{-1}\right)_{ij} = \frac{\det A_i\left(\mathbf{e}_j\right)}{\det A}$$

where $A_i\left(\mathbf{e}_j\right)$ is the matrix obtained by replacing the ith column of A with $\mathbf{e}_j$. To find the $\det A_i\left(\mathbf{e}_j\right)$, we use the cofactor expansion down the ith column of the matrix $A_i\left(\mathbf{e}_j\right)$, which is

$$\det A_i\left(\mathbf{e}_j\right) = (-1)^{i+j} \det A_{ji} = C_{ji}$$

Therefore,

$$\left(A^{-1}\right)_{ij} = \frac{\det A_i\left(\mathbf{e}_j\right)}{\det A} = \frac{C_{ji}}{\det A}$$

which proves that

$$A^{-1} = \frac{1}{\det A}C^T.$$

The A^{-1} is represented as,

$$A^{-1} = \frac{1}{\det A}\begin{bmatrix} C_{11} & C_{12} & \cdots & C_{1n} \\ C_{21} & C_{22} & \cdots & C_{2n} \\ \vdots & \vdots & & \vdots \\ C_{n1} & C_{n2} & \cdots & C_{nn} \end{bmatrix}^T$$

$$= \frac{1}{\det A}\begin{bmatrix} C_{11} & C_{21} & \cdots & C_{n1} \\ C_{12} & C_{22} & \cdots & C_{n2} \\ \vdots & \vdots & & \vdots \\ C_{1n} & C_{2n} & \cdots & C_{nn} \end{bmatrix}$$

The transpose of the cofactor matrix of A is called adjugate of A, denoted by $adj\ A$. Thus

$$A^{-1} = \frac{1}{\det A}C^T = \frac{1}{\det A}adj\ A.$$

Example 4 We use inverse formula to decide the inverse of the matrix

$$A = \begin{bmatrix} 2 & 0 & 0 \\ 1 & -1 & 1 \\ 2 & 3 & -2 \end{bmatrix}$$

The cofactors of A are,

$$C_{11} = \begin{vmatrix} -1 & 1 \\ 3 & -2 \end{vmatrix} = -1, \quad C_{12} = -\begin{vmatrix} 1 & 1 \\ 2 & -2 \end{vmatrix} = 4, \quad C_{13} = \begin{vmatrix} 1 & -1 \\ 2 & 3 \end{vmatrix} = 5$$

$$C_{21} = -\begin{vmatrix} 0 & 0 \\ 3 & -2 \end{vmatrix} = 0, \quad C_{22} = \begin{vmatrix} 2 & 0 \\ 2 & -2 \end{vmatrix} = -4, \quad C_{23} = -\begin{vmatrix} 2 & 0 \\ 2 & 3 \end{vmatrix} = -6$$

$$C_{31} = \begin{vmatrix} 0 & 0 \\ -1 & 1 \end{vmatrix} = 0, \quad C_{32} = -\begin{vmatrix} 2 & 0 \\ 1 & 1 \end{vmatrix} = -2, \quad C_{33} = \begin{vmatrix} 2 & 0 \\ 1 & -1 \end{vmatrix} = -2$$

Thus,

$$Adj\ A = \begin{bmatrix} -1 & 4 & 5 \\ 0 & -4 & -6 \\ 0 & -2 & -2 \end{bmatrix}^T$$

and

$$\begin{aligned} A^{-1} &= \frac{1}{\det A} Adj\ A \\ &= -\frac{1}{2} \begin{bmatrix} -1 & 0 & 0 \\ 4 & -4 & -2 \\ 5 & -6 & -2 \end{bmatrix} \\ &= \begin{bmatrix} \frac{1}{2} & 0 & 0 \\ -2 & 2 & 1 \\ -\frac{5}{2} & 3 & 1 \end{bmatrix}. \end{aligned}$$

Area and Volume

Theorem 5 *A parallelogram defined by two vectors* **u** *and* **v** *in* R^2 *has the area of* $|\det [\mathbf{u}\ \mathbf{v}]|$.

Example 6 Find the area of the parallelogram defined by $\mathbf{u} = \begin{bmatrix} 3 \\ 0 \end{bmatrix}$ and $\mathbf{v} = \begin{bmatrix} 0 \\ -8 \end{bmatrix}$.

Solution *This is a special case. The parallelogram here defined by* **u** *and* **v** *is in fact a rectangle. Its area is*

$$\left| \det \begin{bmatrix} 3 & 0 \\ 0 & -8 \end{bmatrix} \right| = |-24| = 24$$

Example 7 Find the area of the parallelogram defined by $\mathbf{u} = \begin{bmatrix} 1 \\ 3 \end{bmatrix}$ and $\mathbf{v} = \begin{bmatrix} 7 \\ 1 \end{bmatrix}$.

Solution *The area of the parallelogram is*

$$\left|\det\begin{bmatrix}1 & 7\\ 3 & 1\end{bmatrix}\right| = |-20| = 20$$

Theorem 8 *A parallelepiped defined by three vectors* **u**, **v** *and* **w** *in* R^3 *has the volume of* $|\det[\mathbf{u}\ \mathbf{v}\ \mathbf{w}]|$.

Example 9 Find the volume of the parallelepiped defined by $\mathbf{u} = \begin{bmatrix}2\\0\\0\end{bmatrix}$, $\mathbf{v} = \begin{bmatrix}0\\8\\0\end{bmatrix}$, and $\mathbf{w} = \begin{bmatrix}0\\0\\6\end{bmatrix}$.

Solution *This is a special case. The parallelepiped here defined by* **u**, **v**, *and* **w** *is in fact a rectangular box. Its volume is,*

$$\left|\det\begin{bmatrix}2 & 0 & 0\\ 0 & 8 & 0\\ 0 & 0 & 6\end{bmatrix}\right| = |96| = 96$$

Example 10 Find the volume of the parallelepiped defined by $\mathbf{u} = \begin{bmatrix}2\\0\\1\end{bmatrix}$, $\mathbf{v} = \begin{bmatrix}1\\0\\1\end{bmatrix}$, and $\mathbf{w} = \begin{bmatrix}2\\5\\0\end{bmatrix}$.

Solution *The volume of the parallelepiped here defined by* **u**, **v**, *and* **w** *is,*

$$\left|\det\begin{bmatrix}2 & 1 & 2\\ 0 & 0 & 5\\ 1 & 1 & 0\end{bmatrix}\right| = |-5| = 5$$

If a linear transformation $S(\mathbf{x}) = A\mathbf{x}$, where A is a 2×2 matrix, is applied to a region E in R^2, the resulting region $S(E)$ may have an area different than that of E. Similarly, if a linear transformation $T(\mathbf{x}) = B\mathbf{x}$, where B is a 3×3 matrix, is applied to a region F in R^3, the resulting region $T(F)$ may have a volume different than that of F. The following Theorems state how the areas of E and $S(E)$ are related, and as is the volumes of F and $T(F)$.

Theorem 11 *Let $S : R^2 \to R^2$ be a linear transformation defined by $S(\mathbf{x}) = A\mathbf{x}$, where A is a 2×2 matrix. If a region E in R^2 is mapped to the region $S(E)$ in R^2, then*

$$\textit{area of } S(E) = |\det A| \cdot (\textit{area of } E)$$

Theorem 12 *Let $T : R^3 \to R^3$ be a linear transformation defined by $T(\mathbf{x}) = B\mathbf{x}$, where B is a 3×3 matrix. If a region F in R^3 is mapped to the region $T(F)$ in R^3, then*

$$\textit{volume of } T(F) = |\det B| \cdot (\textit{volume of } F).$$

Example 13 Let $S : R^2 \to R^2$ be a linear transformation defined by $S(\mathbf{x}) = A\mathbf{x} = \begin{bmatrix} 2 & 0 \\ 0 & 5 \end{bmatrix} \begin{bmatrix} x_1 \\ x_2 \end{bmatrix}$. Under the transformation of S, a unit disk E is mapped to $S(E)$. Find the area of $S(E)$.

Solution *The unit disk E has a radius of 1. Thus the area of the unit disk is $\pi r^2 = \pi \cdot 1^2 = \pi$. So the image $S(E)$ of the unit disk has the area,*

$$|\det A| \cdot (\textit{area of } E) = 10(\pi) = 10\pi$$

Remark 14 In the above example, the transformation matrix A indicates that S maps $\mathbf{e}_1 = \begin{bmatrix} 1 \\ 0 \end{bmatrix}$ to $\begin{bmatrix} 2 \\ 0 \end{bmatrix}$, and maps $\mathbf{e}_2 = \begin{bmatrix} 0 \\ 1 \end{bmatrix}$ to $\begin{bmatrix} 0 \\ 5 \end{bmatrix}$. The boundary of the unit disk E is in fact mapped to an ellipse. The area of $S(E)$ is the area of the ellipse.

Example 15 Let $T : R^3 \to R^3$ be a linear transformation defined by $T(\mathbf{x}) = B\mathbf{x} = \begin{bmatrix} 2 & 5 & 8 \\ 0 & 1 & 9 \\ 0 & 0 & 6 \end{bmatrix} \begin{bmatrix} x_1 \\ x_2 \\ x_3 \end{bmatrix}$. A parallelepiped F determined by $\mathbf{u} = \begin{bmatrix} 2 \\ 0 \\ 1 \end{bmatrix}$, $\mathbf{v} = \begin{bmatrix} 1 \\ 0 \\ 1 \end{bmatrix}$, and $\mathbf{w} = \begin{bmatrix} 2 \\ 5 \\ 0 \end{bmatrix}$ is mapped to $T(F)$. Find the volume of $T(F)$.

Solution *We know that the volume of the parallelepiped F is*

$$|\det [\mathbf{u}\ \mathbf{v}\ \mathbf{w}]| = \left| \det \begin{bmatrix} 2 & 1 & 2 \\ 0 & 0 & 5 \\ 1 & 1 & 0 \end{bmatrix} \right| = 5.$$

The image $T(F)$ has the volume,

$$|\det B| \cdot (\textit{volume of } F) = 12(5) = 60.$$

4.3 Exercises

Exercise 16 Use Cramer's rule to solve each linear system.

$$\text{(a)}\ \begin{aligned} 2x_1 + x_2 &= 3 \\ x_1 + 5x_2 &= 7 \end{aligned}$$

$$\text{(b)}\ \begin{aligned} x_1 + 2x_2 &= 6 \\ 3x_1 - 4x_2 &= 1 \end{aligned}$$

$$\text{(c)}\ \begin{aligned} x_1 + 3x_2 + x_3 &= -1 \\ 2x_1 + x_2 + x_3 &= 6 \\ -2x_1 + 2x_2 - x_3 &= 1 \end{aligned}$$

$$\text{(d)}\ \begin{bmatrix} 1 & 0 & 0 \\ 0 & 1 & 1 \\ 2 & 0 & 5 \end{bmatrix} \begin{bmatrix} x_1 \\ x_2 \\ x_3 \end{bmatrix} = \begin{bmatrix} 2 \\ 5 \\ 1 \end{bmatrix}$$

Exercise 17 Find adjugate and inverse of each matrix.

$$A = \begin{bmatrix} 1 & 0 & 0 \\ 1 & 5 & 0 \\ 3 & 1 & 2 \end{bmatrix}, \quad B = \begin{bmatrix} 4 & 5 & 1 \\ -2 & -1 & 4 \\ 2 & 1 & -3 \end{bmatrix}, \quad C = \begin{bmatrix} 2 & 1 & 5 \\ 1 & 3 & 5 \\ -2 & 2 & -5 \end{bmatrix}$$

Exercise 18 Show that if B is invertible then the matrix $adj\ B$ is invertible, and

$$(adj\ B)^{-1} = \det\left(B^{-1}\right) B.$$

Exercise 19 Find the area of the parallelogram defined by $\mathbf{u} = \begin{bmatrix} 6 \\ -4 \end{bmatrix}$ and $\mathbf{v} = \begin{bmatrix} 1 \\ 2 \end{bmatrix}$.

Exercise 20 Find the volume of the parallelepiped defined by $\mathbf{u} = \begin{bmatrix} 1 \\ 0 \\ 3 \end{bmatrix}$, $\mathbf{v} = \begin{bmatrix} 2 \\ 2 \\ 0 \end{bmatrix}$, and $\mathbf{w} = \begin{bmatrix} 1 \\ 0 \\ 6 \end{bmatrix}$.

Exercise 21 Let $S : R^2 \to R^2$ be a linear transformation defined by $S(\mathbf{x}) = A\mathbf{x} = \begin{bmatrix} -3 & 1 \\ 5 & 7 \end{bmatrix} \begin{bmatrix} x_1 \\ x_2 \end{bmatrix}$. A parallelogram E determined by $\mathbf{u} = \begin{bmatrix} 1 \\ 3 \end{bmatrix}$ and $\mathbf{v} = \begin{bmatrix} 5 \\ 9 \end{bmatrix}$ is mapped to $S(E)$. Find the area of $S(E)$.

Exercise 22 Let $T : R^3 \to R^3$ be a linear transformation defined by $T(\mathbf{x}) = B\mathbf{x} = \begin{bmatrix} 1 & 0 & 1 \\ 2 & 4 & 6 \\ 2 & 1 & 0 \end{bmatrix} \begin{bmatrix} x_1 \\ x_2 \\ x_3 \end{bmatrix}$. A parallelepiped F determined by $\mathbf{u} = \begin{bmatrix} 2 \\ 0 \\ 1 \end{bmatrix}$, $\mathbf{v} = \begin{bmatrix} 1 \\ 0 \\ 1 \end{bmatrix}$, and $\mathbf{w} = \begin{bmatrix} 2 \\ 5 \\ 0 \end{bmatrix}$ is mapped to $T(F)$. Find the volume of $T(F)$.

Exercise 23 Let $S : R^2 \to R^2$ be a linear transformation defined by $S(\mathbf{x}) = A\mathbf{x} = \begin{bmatrix} 5 & 0 \\ 0 & -4 \end{bmatrix} \begin{bmatrix} x_1 \\ x_2 \end{bmatrix}$. A unit disk E is mapped to $S(E)$. Find the area of $S(E)$.

Exercise 24 Let $T : R^3 \to R^3$ be a linear transformation defined by $T(\mathbf{x}) = B\mathbf{x} = \begin{bmatrix} 1 & 0 & 0 \\ 0 & 2 & 0 \\ 0 & 0 & 6 \end{bmatrix} \begin{bmatrix} x_1 \\ x_2 \\ x_3 \end{bmatrix}$. A unit ball F is mapped to $T(F)$. Find the volume of $T(F)$.

5 Eigenvalues and Eigenvectors

5.1 Eigenvalues and Eigenvectors

We consider a linear transformation defined by an $n \times n$ matrix A. If there is a nonzero vector $\mathbf{v}$ such that $A\mathbf{v} = \lambda\mathbf{v}$ for some scalar λ, then the scalar λ is called an **eigenvalue** of A, and the vector $\mathbf{v}$ is called an **eigenvector** corresponding to λ.

Example 1 Let

$$A = \begin{bmatrix} 2 & 1 \\ 0 & -1 \end{bmatrix}, \quad \mathbf{u} = \begin{bmatrix} 5 \\ 0 \end{bmatrix}, \quad \mathbf{v} = \begin{bmatrix} 1 \\ 5 \end{bmatrix}.$$

We decide if $\mathbf{u}$ and $\mathbf{v}$ are eigenvectors of A. Since

$$A\mathbf{u} = \begin{bmatrix} 2 & 1 \\ 0 & -1 \end{bmatrix}\begin{bmatrix} 5 \\ 0 \end{bmatrix} = \begin{bmatrix} 10 \\ 0 \end{bmatrix} = 2\begin{bmatrix} 5 \\ 0 \end{bmatrix} = 2\mathbf{u},$$

it follows that $\lambda = 2$ is an eigenvalue of A and $\mathbf{u} = \begin{bmatrix} 5 \\ 0 \end{bmatrix}$ is an eigenvector corresponding to $\lambda = 2$. Since

$$A\mathbf{v} = \begin{bmatrix} 2 & 1 \\ 0 & -1 \end{bmatrix}\begin{bmatrix} 1 \\ 5 \end{bmatrix} = \begin{bmatrix} 7 \\ -5 \end{bmatrix} \neq \lambda\begin{bmatrix} 1 \\ 5 \end{bmatrix} \text{ for any number } \lambda,$$

$\mathbf{v}$ is not an eigenvector of A.

Remark 2 If $\mathbf{v}$ is an eigenvector of A, the image of $\mathbf{v}$ under the linear transformation of A is simply a scalar multiple of $\mathbf{v}$.

H. Tian, *Linear Algebra*, Synthesis Lectures on Mathematics & Statistics,
https://doi.org/10.1007/978-3-031-84647-2_5

Example 3 Decide whether $\lambda = 4$ is an eigenvalue of $A = \begin{bmatrix} 3 & 2 \\ 3 & -2 \end{bmatrix}$.

Solution *We need to decide if there exists an eigenvector* $\mathbf{v}$ *corresponding to* $\lambda = 4$. *That is, we need to decide if the equation* $A\mathbf{v} = 4\mathbf{v}$ *has at least a nonzero solution* $\mathbf{v}$. *Now we consider the system,*

$$\begin{bmatrix} 3 & 2 \\ 3 & -2 \end{bmatrix} \begin{bmatrix} v_1 \\ v_2 \end{bmatrix} = 4 \begin{bmatrix} v_1 \\ v_2 \end{bmatrix}$$

equivalently,

$$\begin{aligned} 3v_1 + 2v_2 &= 4v_1 \\ 3v_1 - 2v_2 &= 4v_2 \end{aligned}$$

which is in fact a homogeneous system

$$\begin{aligned} -v_1 + 2v_2 &= 0 \\ 3v_1 - 6v_2 &= 0 \end{aligned}$$

equivalently,

$$\begin{bmatrix} -1 & 2 \\ 3 & -6 \end{bmatrix} \begin{bmatrix} v_1 \\ v_2 \end{bmatrix} = \begin{bmatrix} 0 \\ 0 \end{bmatrix}.$$

The row echelon form of the coefficient matrix

$$\begin{bmatrix} -1 & 2 \\ 3 & -6 \end{bmatrix} \sim \begin{bmatrix} -1 & 2 \\ 0 & 0 \end{bmatrix} \sim \begin{bmatrix} 1 & -2 \\ 0 & 0 \end{bmatrix}$$

indicates a free variable. Therefore there exists at least a nonzero solution for $A\mathbf{v} = 4\mathbf{v}$. *So 4 is an eigenvalue of* A. *Note that* $A\mathbf{v} = 4\mathbf{v}$ *has the general solutions* $\mathbf{v} = \begin{bmatrix} 2v_2 \\ v_2 \end{bmatrix} = v_2 \begin{bmatrix} 2 \\ 1 \end{bmatrix}$.

That is, any nonzero vector $\mathbf{v}$ *as a scalar multiple of* $\begin{bmatrix} 2 \\ 1 \end{bmatrix}$ *is an eigenvector corresponding to* $\lambda = 4$.

Example 4 Find eigenvalues of $A = \begin{bmatrix} 2 & 0 \\ -2 & 6 \end{bmatrix}$.

Solution *A number* λ *is an eigenvalue of* A *if and only if there exists a nontrivial solution for* $A\mathbf{v} = \lambda\mathbf{v}$, *i.e.,*

$$\begin{bmatrix} 2 & 0 \\ -2 & 6 \end{bmatrix} \begin{bmatrix} v_1 \\ v_2 \end{bmatrix} = \lambda \begin{bmatrix} v_1 \\ v_2 \end{bmatrix}$$

which can be written as a homogeneous system as follows,

$$\begin{bmatrix} 2-\lambda & 0 \\ -2 & 6-\lambda \end{bmatrix} \begin{bmatrix} v_1 \\ v_2 \end{bmatrix} = \begin{bmatrix} 0 \\ 0 \end{bmatrix}.$$

The system has a nontrivial solution if and only if its coefficient matrix $\begin{bmatrix} 2-\lambda & 0 \\ -2 & 6-\lambda \end{bmatrix}$ *is singular, i.e.,*

$$\det \begin{bmatrix} 2-\lambda & 0 \\ -2 & 6-\lambda \end{bmatrix} = 0.$$

Otherwise, if $\det \begin{bmatrix} 2-\lambda & 0 \\ -2 & 6-\lambda \end{bmatrix} \neq 0$, *the homogeneous system* $\begin{bmatrix} 2-\lambda & 0 \\ -2 & 6-\lambda \end{bmatrix} \begin{bmatrix} v_1 \\ v_2 \end{bmatrix} = \begin{bmatrix} 0 \\ 0 \end{bmatrix}$ *would have only a trivial solution. Solving* $\det \begin{bmatrix} 2-\lambda & 0 \\ -2 & 6-\lambda \end{bmatrix} = 0$, *that is,*

$$(2-\lambda)(6-\lambda) = 0$$

We obtain the eigenvalues of A as $\lambda = 2$ *and* $\lambda = 6$.

A number λ is an eigenvalue of an $n \times n$ matrix A if and only if one of the following statements is true. (The I in the table below denotes an $n \times n$ identity matrix).

(a) The equation $A\mathbf{v} = \lambda\mathbf{v}$, i.e., $(A - \lambda I)\mathbf{v} = \mathbf{0}$, has a nontrivial solution.
(b) $Nul\ (A - \lambda I) \neq \{\mathbf{0}\}$.
(c) the matrix $A - \lambda I$ is not invertible
(d) $\det(A - \lambda I) = 0$.

The eigenvectors $\mathbf{v}$ corresponding to λ satisfy the equation $(A - \lambda I)\mathbf{v} = \mathbf{0}$. The subspace $Nul\ (A - \lambda I)$ is called the **eigenspace** corresponding to the eigenvalue λ. It is a subspace that includes not only all the eigenvectors corresponding to λ, but also the $\mathbf{0}$ vector.

The equation $\det(A - \lambda I) = 0$ is called the **characteristic equation** of A. The polynomial $p(\lambda) = \det(A - \lambda I)$ is said to be the **characteristic polynomial** of A. More specifically,

$$p(\lambda) = \det(A - \lambda I) = \begin{vmatrix} a_{11}-\lambda & a_{12} & \cdots & a_{1n} \\ a_{21} & a_{22}-\lambda & \cdots & a_{2n} \\ \vdots & & \ddots & \vdots \\ a_{n1} & a_{n2} & \cdots & a_{nn}-\lambda \end{vmatrix}$$

As we already know, finding the determinant of a dense matrix may involve a lot of calculation. The task is relatively easier when a matrix is sparse, i.e., when a matrix has many zero entries.

Example 5 Find the eigenvalues of the triangular matrix $A = \begin{bmatrix} 2 & 0 & 5 & 2 \\ 0 & 2 & 7 & -3 \\ 0 & 0 & 0 & 4 \\ 0 & 0 & 0 & 6 \end{bmatrix}$.

Solution *A number* λ *is an eigenvalue of A if and only if there exists a nontrivial solution for the equation* $A\mathbf{v} = \lambda\mathbf{v}$,

$$\begin{bmatrix} 2 & 0 & 5 & 2 \\ 0 & 2 & 7 & -3 \\ 0 & 0 & 0 & 4 \\ 0 & 0 & 0 & 6 \end{bmatrix} \begin{bmatrix} v_1 \\ v_2 \\ v_3 \\ v_4 \end{bmatrix} = \lambda \begin{bmatrix} v_1 \\ v_2 \\ v_3 \\ v_4 \end{bmatrix}$$

which is the same as the homogeneous system below,

$$\begin{bmatrix} 2-\lambda & 0 & 5 & 2 \\ 0 & 2-\lambda & 7 & -3 \\ 0 & 0 & 0-\lambda & 4 \\ 0 & 0 & 0 & 6-\lambda \end{bmatrix} \begin{bmatrix} v_1 \\ v_2 \\ v_3 \\ v_4 \end{bmatrix} = \begin{bmatrix} 0 \\ 0 \\ 0 \\ 0 \end{bmatrix}.$$

The homogeneous system has a nontrivial solution if and only if its coefficient matrix is singular, i.e.,

$$\det \begin{bmatrix} 2-\lambda & 0 & 5 & 2 \\ 0 & 2-\lambda & 7 & -3 \\ 0 & 0 & 0-\lambda & 4 \\ 0 & 0 & 0 & 6-\lambda \end{bmatrix} = 0.$$

Recall that the determinant of a triangular matrix is the product of its diagonal entries. So :

$$(2-\lambda)^2 (0-\lambda)(6-\lambda) = 0.$$

Therefore, the eigenvalues of the matrix A are 2, 0, *and* 6.

Remark 6 Eigenvectors are nonzero vectors. But eigenvalues can be 0.

Theorem 7 *The eigenvalues of a triangular matrix are the values on the main diagonal.*

Proof Let A be an $n \times n$ triangular matrix,

$$A = \begin{bmatrix} a_{11} & a_{12} & \cdots & a_{1n} \\ 0 & a_{22} & \cdots & a_{2n} \\ 0 & 0 & \ddots & \vdots \\ 0 & 0 & 0 & a_{nn} \end{bmatrix}.$$

A number λ is an eigenvalue of A if and only if there exists a nontrivial solution for $(A - \lambda I)\,\mathbf{v} = \mathbf{0}$, which is,

$$\begin{bmatrix} a_{11}-\lambda & a_{12} & \cdots & a_{1n} \\ 0 & a_{22}-\lambda & \cdots & a_{2n} \\ 0 & 0 & \ddots & \vdots \\ 0 & 0 & 0 & a_{nn}-\lambda \end{bmatrix} \begin{bmatrix} v_1 \\ v_2 \\ \vdots \\ v_4 \end{bmatrix} = \begin{bmatrix} 0 \\ 0 \\ 0 \\ 0 \end{bmatrix}.$$

Therefore the coefficient matrix must be singular, i.e.,

$$\det \begin{bmatrix} a_{11}-\lambda & a_{12} & \cdots & a_{1n} \\ 0 & a_{22}-\lambda & \cdots & a_{2n} \\ 0 & 0 & \ddots & \vdots \\ 0 & 0 & 0 & a_{nn}-\lambda \end{bmatrix} = 0.$$

Thus,

$$(a_{11}-\lambda)\,(a_{22}-\lambda)\cdots(a_{nn}-\lambda) = 0.$$

Therefore, the eigenvalues of a triangular matrix are the values on its main diagonal: a_{11}, $a_{22}, \ldots, a_{nn}$. ■

To find eigenvalues of any matrix A, we need to solve the characteristic equation $\det\,[A - \lambda I] = 0$ for λ.

Example 8 Find the eigenvalues and the corresponding eigenspaces for the matrix $A = \begin{bmatrix} 1 & 2 & 1 \\ 0 & 3 & 1 \\ 0 & 5 & -1 \end{bmatrix}$.

Solution *First, we find eigenvalues of A by solving its characteristic equation* $\det\,[A - \lambda I] = 0$.

$$\begin{vmatrix} 1-\lambda & 2 & 1 \\ 0 & 3-\lambda & 1 \\ 0 & 5 & -1-\lambda \end{vmatrix} = 0$$

Using cofactor expansion down the first column, the characteristic equation can be reduced to,

$$(1-\lambda)\begin{vmatrix} 3-\lambda & 1 \\ 5 & -1-\lambda \end{vmatrix} = 0$$

which is

$$(1-\lambda)\,[(3-\lambda)\,(-1-\lambda) - 5] = 0$$

i.e.,

$$(1-\lambda)\left[\lambda^2-2\lambda-8\right]=0$$

Hence,

$$(1-\lambda)(\lambda-4)(\lambda+2)=0.$$

So the eigenvalues of A are 1, 4, *and* −2. *Next, we find the eigenspace* E_λ *corresponding to each eigenvalue* λ *by solving* $(A-\lambda I)\,\mathbf{v}=\mathbf{0}$.

$$For\ \lambda=1,$$

$$\left[\begin{bmatrix}0&2&1\\0&2&1\\0&5&-2\end{bmatrix}\right]\begin{bmatrix}v_1\\v_2\\v_3\end{bmatrix}=\mathbf{0},$$

$$E_1=Span\left\{\begin{bmatrix}1\\0\\0\end{bmatrix}\right\}$$

$$For\ \lambda=4,$$

$$\begin{bmatrix}-3&2&1\\0&-1&1\\0&5&-5\end{bmatrix}\begin{bmatrix}v_1\\v_2\\v_3\end{bmatrix}=\mathbf{0},$$

$$E_4=Span\left\{\begin{bmatrix}1\\1\\1\end{bmatrix}\right\}$$

$$For\ \lambda=-2,$$

$$\begin{bmatrix}3&2&1\\0&5&1\\0&5&1\end{bmatrix}\begin{bmatrix}v_1\\v_2\\v_3\end{bmatrix}=\mathbf{0},$$

$$E_{-2}=Span\left\{\begin{bmatrix}-1\\-1\\5\end{bmatrix}\right\}.$$

5.1 Exercises

Exercise 9 Let $A = \begin{bmatrix} 1 & 1 \\ 3 & -1 \end{bmatrix}$. Decide if $\mathbf{u} = \begin{bmatrix} 2 \\ 5 \end{bmatrix}$ is an eigenvector of A.

Exercise 10 Decide whether $\lambda = 2$ is an eigenvalue of $A = \begin{bmatrix} 3 & -1 \\ 1 & 1 \end{bmatrix}$. If it is an eigenvalue, find its corresponding eigenvectors.

Exercise 11 Find the eigenvalues of each triangular matrix.

$$A = \begin{bmatrix} 5 & 0 & 0 \\ 2 & 11 & 0 \\ 6 & 4 & -3 \end{bmatrix}, \quad B = \begin{bmatrix} 7 & 1 & 2 & 6 \\ 0 & 3 & -4 & 2 \\ 0 & 0 & 3 & 8 \\ 0 & 0 & 0 & 1 \end{bmatrix}, \quad C = \begin{bmatrix} 5 & 0 & 0 & 0 \\ 1 & -1 & 0 & 0 \\ 0 & -1 & 4 & 0 \\ 7 & 0 & 3 & 8 \end{bmatrix}$$

Exercise 12 Find the eigenvalues and the corresponding eigenspaces for each matrix.

$$A = \begin{bmatrix} 1 & -2 \\ -2 & 1 \end{bmatrix}, \quad B = \begin{bmatrix} 0 & 1 & 0 \\ 0 & 0 & 1 \\ 0 & 0 & 0 \end{bmatrix}, \quad C = \begin{bmatrix} 1 & 1 & 1 \\ 0 & 2 & 1 \\ 0 & 0 & 1 \end{bmatrix}, \quad D = \begin{bmatrix} 2 & 5 & 11 \\ 0 & 1 & 3 \\ 0 & 0 & 6 \end{bmatrix}$$

Exercise 13 Find the eigenvalues and the corresponding eigenspaces for each matrix

$$A = \begin{bmatrix} 5 & 1 & 2 \\ 0 & -1 & 2 \\ 0 & -3 & 4 \end{bmatrix}, \quad B = \begin{bmatrix} 2 & 0 & 0 & 0 \\ 5 & 1 & 0 & 0 \\ 0 & 0 & 3 & 1 \\ 0 & 0 & 0 & 3 \end{bmatrix}.$$

5.2 Properties of Eigenvalues and Eigenvectors

We can determine if a square matrix is invertible based on the eigenvalues of the matrix.

Example 1 Decide if the matrix $A = \begin{bmatrix} 2 & 0 & 5 & 2 \\ 0 & 2 & 7 & -3 \\ 0 & 0 & 0 & 4 \\ 0 & 0 & 0 & 6 \end{bmatrix}$ is invertible.

Solution *Solving the characteristic equation* $\det[A - \lambda I] = 0$, *we know A has eigenvalues* $\lambda = 2, 0,$ *and* 6. *Note that A has an eigenvalue* $\lambda = 0$. *This implies* $\det[A - 0I] = 0$, *i.e.,* $\det A = 0$. *So when A has a* 0 *eigenvalue, the matrix A must be singular and not invertible.*

On the other hand, if A is singular, i.e., $\det A = 0$, *which is the same as* $\det[A - 0I] = 0$, *then* 0 *is an eigenvalue.*

Theorem 2 *An* $n \times n$ *matrix A is invertible if and only if* 0 *is not an eigenvalue.*

Example 3 The matrix $A = \begin{bmatrix} 3 & 2 \\ 3 & -2 \end{bmatrix}$ has the characteristic equation

$$\begin{vmatrix} 3-\lambda & 2 \\ 3 & -2-\lambda \end{vmatrix} = 0$$

i.e.,

$$\lambda^2 - \lambda - 12 = 0$$

So A has eigenvalues $\lambda = 4, -3$. None of the eigenvalues is 0. So the matrix A is invertible

Next, let's review an example of $n \times n$ matrix that has n distinct eigenvalues.

Example 4 From previous section, the 3×3 matrix $A = \begin{bmatrix} 1 & 2 & 1 \\ 0 & 3 & 1 \\ 0 & 5 & -1 \end{bmatrix}$ has 3 distinct eigenvalues $\lambda = 1$, 4, and -2. The eigenspace E_λ corresponding to each eigenvalue are respectively

$$E_1 = Span\left\{\begin{bmatrix} 1 \\ 0 \\ 0 \end{bmatrix}\right\}, \quad E_4 = Span\left\{\begin{bmatrix} 1 \\ 1 \\ 1 \end{bmatrix}\right\}, \quad E_{-2} = Span\left\{\begin{bmatrix} -1 \\ -1 \\ 5 \end{bmatrix}\right\}.$$

We see that the set $\left\{\begin{bmatrix} 1 \\ 0 \\ 0 \end{bmatrix}, \begin{bmatrix} 1 \\ 1 \\ 1 \end{bmatrix}, \begin{bmatrix} -1 \\ -1 \\ 5 \end{bmatrix}\right\}$ of eigenvectors corresponding to 3 different eigenvalues is linearly independent. It is not hard to check on the linear independence of any 3 eigenvectors belonging to distinct eigenvalues of A.

It is in fact true that if an $n \times n$ matrix A has n disctinct eigenvalues, then the eigenvectors belonging to disctinct eigenvalues form a linearly independent set. The following Theorem gives a stronger result.

Theorem 5 *Let A be an* $n \times n$ *matrix. Suppose A has k distinct eigenvalues* $\lambda_1, \lambda_2, \ldots, \lambda_k$ *and the vectors* $\mathbf{v}_1, \mathbf{v}_2, \ldots, \mathbf{v}_k$ *are eigenvectors corresponding to the distinct eigenvalues. Then the set* $\{\mathbf{v}_1, \mathbf{v}_2, \ldots, \mathbf{v}_k\}$ *is linearly independent.*

Proof We consider the largest possible m, $(m \leq k)$, such that the set $\{\mathbf{v}_1, \mathbf{v}_2, \ldots, \mathbf{v}_m\}$ is linearly independent. We want to show that $m = k$. If suppose $m < k$, then $\{\mathbf{v}_1, \mathbf{v}_2, \ldots, \mathbf{v}_m, \mathbf{v}_{m+1}\}$ would be a linearly dependent set and the vector $\mathbf{v}_{m+1}$ can be written as a linear combination of the preceding vectors $\mathbf{v}_1, \mathbf{v}_2, \ldots, \mathbf{v}_m$. Thus, there exists $x_1, x_2, \ldots,$ and x_m, not all zeroes, such that,

$$\mathbf{v}_{m+1} = x_1\mathbf{v}_1 + x_2\mathbf{v}_2 + \cdots + x_m\mathbf{v}_m$$

Applying the linear transformation defined by A onto both sides of the above equation, we obtain

$$A\mathbf{v}_{m+1} = x_1 A\mathbf{v}_1 + x_2 A\mathbf{v}_2 + \cdots + x_m A\mathbf{v}_m$$

Using the fact that $\{\mathbf{v}_1, \mathbf{v}_2, \ldots, \mathbf{v}_k\}$ are eigenvectors belonging respectively to $\lambda_1, \lambda_2, \ldots, \lambda_k$,

$$\lambda_{m+1}\mathbf{v}_{m+1} = x_1\lambda_1\mathbf{v}_1 + x_2\lambda_2\mathbf{v}_2 + \cdots + x_m\lambda_m\mathbf{v}_m$$

However, the equation $\mathbf{v}_{m+1} = x_1\mathbf{v}_1 + x_2\mathbf{v}_2 + \cdots + x_m\mathbf{v}_m$ also implies the following,

$$\lambda_{m+1}\mathbf{v}_{m+1} = x_1\lambda_{m+1}\mathbf{v}_1 + x_2\lambda_{m+1}\mathbf{v}_2 + \cdots + x_m\lambda_{m+1}\mathbf{v}_m$$

Subtracting the two expressions for $\lambda_{m+1}\mathbf{v}_{m+1}$, we have

$$\mathbf{0} = x_1(\lambda_1 - \lambda_{m+1})\,\mathbf{v}_1 + x_2(\lambda_2 - \lambda_{m+1})\,\mathbf{v}_2 + \cdots + x_m(\lambda_m - \lambda_{m+1})\,\mathbf{v}_m$$

Since the set $\{\mathbf{v}_1, \mathbf{v}_2, \ldots, \mathbf{v}_m\}$ is linearly independent, the weights of the vectors in the above equation must all be zeros, $\because$

$$\begin{aligned} x_1(\lambda_1 - \lambda_{m+1}) &= 0, \\ x_2(\lambda_2 - \lambda_{m+1}) &= 0, \\ &\vdots \\ x_m(\lambda_m - \lambda_{m+1}) &= 0. \end{aligned}$$

Knowing that $\lambda_1, \lambda_2, \ldots, \lambda_k$ are distinct eigenvalues, thus,

$$\begin{aligned} x_1 &= 0, \\ x_2 &= 0, \\ &\vdots \\ x_m &= 0 \end{aligned}$$

This is contradictory to our assumption. Therefore $m = k$. ■

Corollary 6 *Let A be an $n \times n$ matrix that has n distinct eigenvalues $\lambda_1, \lambda_2, \ldots, \lambda_n$ and their corresponding eigenvectors vectors $\mathbf{v}_1, \mathbf{v}_2, \ldots, \mathbf{v}_n$. If P is the $n \times n$ matrix whose columns are the eigenvectors $\mathbf{v}_1, \mathbf{v}_2, \ldots, \mathbf{v}_n$, i.e., $P = [\mathbf{v}_1\ \mathbf{v}_2\ \cdots\ \mathbf{v}_n]$, and* $\Lambda = \begin{bmatrix} \lambda_1 & 0 & 0 & 0 \\ 0 & \lambda_2 & 0 & 0 \\ \vdots & \vdots & \ddots & \vdots \\ 0 & 0 & 0 & \lambda_n \end{bmatrix}$, *then $P^{-1}AP = \Lambda$.*

Proof Since

$$\begin{aligned} A\mathbf{v}_1 &= \lambda_1 \mathbf{v}_1 \\ A\mathbf{v}_2 &= \lambda_2 \mathbf{v}_2 \\ &\vdots \\ A\mathbf{v}_n &= \lambda_n \mathbf{v}_n \end{aligned}$$

We have,

$$\begin{aligned} AP &= A\,[\mathbf{v}_1\ \mathbf{v}_2\ \cdots\ \mathbf{v}_n] \\ &= [A\mathbf{v}_1\ A\mathbf{v}_2\ \cdots\ A\mathbf{v}_n] \\ &= [\lambda_1\mathbf{v}_1\ \lambda_2\mathbf{v}_2\ \cdots\ \lambda_n\mathbf{v}_n] \\ &= [\mathbf{v}_1\ \mathbf{v}_2\ \cdots\ \mathbf{v}_n] \begin{bmatrix} \lambda_1 & 0 & 0 & 0 \\ 0 & \lambda_2 & 0 & 0 \\ \vdots & \vdots & \ddots & \vdots \\ 0 & 0 & 0 & \lambda_n \end{bmatrix} \\ &= P\Lambda \end{aligned}$$

Therefore,

$$AP = P\Lambda$$

The vectors $\mathbf{v}_1, \mathbf{v}_2, \ldots, \mathbf{v}_n$ are eigenvectors belonging to distinct eigenvalues, we know they are independent. So the matrix $P = [\mathbf{v}_1\ \mathbf{v}_2\ \cdots\ \mathbf{v}_n]$ is invertible. We multiply both sides of the equation $AP = P\Lambda$ by P^{-1} to obtain

$$\begin{aligned} P^{-1}AP &= P^{-1}P\Lambda \\ &= \Lambda \end{aligned}$$

■

Example 7 Consider the same matrix $A = \begin{bmatrix} 1 & 2 & 1 \\ 0 & 3 & 1 \\ 0 & 5 & -1 \end{bmatrix}$ as above. The three eigenvectors

$$\mathbf{v}_1 = \begin{bmatrix} 1 \\ 0 \\ 0 \end{bmatrix}, \ \mathbf{v}_2 = \begin{bmatrix} 1 \\ 1 \\ 1 \end{bmatrix}, \ \mathbf{v}_3 = \begin{bmatrix} -1 \\ -1 \\ 5 \end{bmatrix}$$

correspond to eigenvalues $\lambda = 1, 4$, and -2, respectively. If P is the matrix whose 3 columns are $\mathbf{v}_1$, $\mathbf{v}_2$, and $\mathbf{v}_3$, i.e.,

$$P = \begin{bmatrix} 1 & 1 & -1 \\ 0 & 1 & -1 \\ 0 & 1 & 5 \end{bmatrix}$$

then

$$P^{-1}AP = \begin{bmatrix} 1 & 1 & -1 \\ 0 & 1 & -1 \\ 0 & 1 & 5 \end{bmatrix}^{-1} \begin{bmatrix} 1 & 2 & 1 \\ 0 & 3 & 1 \\ 0 & 5 & -1 \end{bmatrix} \begin{bmatrix} 1 & 1 & -1 \\ 0 & 1 & -1 \\ 0 & 1 & 5 \end{bmatrix} = \begin{bmatrix} 1 & 0 & 0 \\ 0 & 4 & 0 \\ 0 & 0 & -2 \end{bmatrix}.$$

Suppose A and B are both $n \times n$ matrices, A is said to be **similar** to B if there exists an invertible matrix P such that $P^{-1}AP = B$.

If A is similar to B, then $P^{-1}AP = B$, which can be written as $A = PBP^{-1}$. Substituting Q for P^{-1}, we have $A = Q^{-1}BQ$. So B is similar to A whenever A is similar to B. We can say A and B are similar.

Theorem 8 *Let A and B are both $n \times n$ matrices, and A is similar to B. Then A and B have the same characteristic equation and the same eigenvalues.*

Proof Since A and B are similar, there exists an invertible matrix P such that

$$P^{-1}AP = B.$$

The characteristic polynomial for B is the same as that for A, since

$$\begin{aligned} \det[B - \lambda I] &= \det\left[P^{-1}AP - \lambda I\right] \\ &= \det\left[P^{-1}AP - \lambda P^{-1}P\right] \\ &= \det P^{-1}[A - \lambda I]\,P \\ &= \det P^{-1} \det P \det[A - \lambda I] \\ &= \det\left(P^{-1}P\right)\det[A - \lambda I] \\ &= \det I \det[A - \lambda I] \\ &= \det[A - \lambda I] \end{aligned}$$

Hence, A and B have the same characteristic equation. So A and B have the same eigenvalues. ■

Remark 9 According to the above Theorem, if A and B are similar, then they have the same eigenvalues with the same multiplicities. However, if A and B have the same eigenvalues, it is not necessarily true that A and B are similar.

Example 10 The two matrices $A = \begin{bmatrix} 3 & 1 \\ 0 & 3 \end{bmatrix}$ and $B = \begin{bmatrix} 3 & 0 \\ 0 & 3 \end{bmatrix}$ have the same eigenvalues, but they are not similar.

The matrix A is said to be **diagonalizable** if A is similar to a diagonal matrix Λ. We know that a matrix A with n distinct eigenvalues is diagonalizable. But what if A has less than n distinct eigenvalues? The following Theorem gives a necessary and sufficient condition for A to be a diagonalizable matrix.

Theorem 11 *An $n \times n$ matrix A is diagonalizable if and only if A has n independent eigenvectors.*

Example 12 Decide if $A = \begin{bmatrix} 2 & 2 & 2 \\ 0 & 0 & 0 \\ 0 & 0 & 0 \end{bmatrix}$ is diagonalizable.

Solution *The 3×3 matrix A is diagonalizable if and only it has 3 independent eigenvectors. Solving the characteristic equation*

$$\begin{vmatrix} 2-\lambda & 2 & 2 \\ 0 & 0-\lambda & 0 \\ 0 & 0 & 0-\lambda \end{vmatrix} = 0$$

which is $\lambda^2 (2 - \lambda) = 0$, we know the matrix has 2 distinct eigenvalues $\lambda = 0$ and $\lambda = 2$. The eigenvectors corresponding to $\lambda = 0$ satisfy the equation

$$\begin{bmatrix} 2 & 2 & 2 \\ 0 & 0 & 0 \\ 0 & 0 & 0 \end{bmatrix} \begin{bmatrix} v_1 \\ v_2 \\ v_3 \end{bmatrix} = \begin{bmatrix} 0 \\ 0 \\ 0 \end{bmatrix}$$

We obtain a basis for the eigenspace E_0 as: $\left\{ \begin{bmatrix} -1 \\ 1 \\ 0 \end{bmatrix}, \begin{bmatrix} -1 \\ 0 \\ 1 \end{bmatrix} \right\}$. *The eigenvectors corresponding to $\lambda = 2$ satisfy the equation*

$$\begin{bmatrix} 0 & 2 & 2 \\ 0 & -2 & 0 \\ 0 & 0 & -2 \end{bmatrix} \begin{bmatrix} v_1 \\ v_2 \\ v_3 \end{bmatrix} = \begin{bmatrix} 0 \\ 0 \\ 0 \end{bmatrix}$$

We obtain a basis for the eigenspace E_2 *as:* $\left\{ \begin{bmatrix} 1 \\ 0 \\ 0 \end{bmatrix} \right\}$. *The* 3×3 *matrix* A *has* 3 *independent eigenvectors* $\left\{ \begin{bmatrix} -1 \\ 1 \\ 0 \end{bmatrix}, \begin{bmatrix} -1 \\ 0 \\ 1 \end{bmatrix}, \begin{bmatrix} 1 \\ 0 \\ 0 \end{bmatrix} \right\}$. *Hence* A *is diagonalizable. If we let* $P = \begin{bmatrix} -1 & -1 & 1 \\ 1 & 0 & 0 \\ 0 & 1 & 0 \end{bmatrix}$, *then*

$$P^{-1}AP = \begin{bmatrix} 0 & 0 & 0 \\ 0 & 0 & 0 \\ 0 & 0 & 2 \end{bmatrix}.$$

5.2 Exercises

Exercise 13 Decide if each triangular matrix is invertible.

$$A = \begin{bmatrix} 5 & 0 & 0 \\ 2 & 11 & 0 \\ 6 & 4 & -3 \end{bmatrix}, \quad B = \begin{bmatrix} 7 & 1 & 2 & 6 \\ 0 & 3 & -4 & 2 \\ 0 & 0 & 3 & 8 \\ 0 & 0 & 0 & 1 \end{bmatrix}, \quad C = \begin{bmatrix} 5 & 0 & 0 & 0 \\ 1 & 0 & 0 & 0 \\ 0 & -1 & 4 & 0 \\ 7 & 0 & 3 & 8 \end{bmatrix}$$

Exercise 14 Diagonalize each matrix if possible.

$$A = \begin{bmatrix} 1 & -2 \\ -2 & 1 \end{bmatrix}, \quad B = \begin{bmatrix} 0 & 1 & 0 \\ 0 & 0 & 1 \\ 0 & 0 & 0 \end{bmatrix}, \quad C = \begin{bmatrix} 2 & 1 & 0 \\ 0 & 2 & 0 \\ 0 & 0 & 3 \end{bmatrix}, \quad D = \begin{bmatrix} 1 & 2 & 2 \\ 0 & 2 & 1 \\ 0 & 0 & 5 \end{bmatrix}$$

Exercise 15 Diagonalize each matrix if possible.

$$A = \begin{bmatrix} 5 & 1 & 2 \\ 0 & -1 & 2 \\ 0 & -3 & 4 \end{bmatrix}, \quad B = \begin{bmatrix} 2 & 0 & 0 & 0 \\ 5 & 1 & 0 & 0 \\ 0 & 0 & 3 & 1 \\ 0 & 0 & 0 & 3 \end{bmatrix}.$$

Orthogonality 6

6.1 Inner Product Spaces

An **inner product** on a vector space V is an operation on V that assigns to each pair of vectors $\mathbf{u}$ and $\mathbf{v}$ in V a real number $(\mathbf{u}, \mathbf{v})$, satisfying the following conditions:

(i) $(\mathbf{u}, \mathbf{u}) \geq 0$, $(\mathbf{u}, \mathbf{u}) = 0$ if and only if $\mathbf{u} = \mathbf{0}$.
(ii) $(\mathbf{u}, \mathbf{v}) = (\mathbf{v}, \mathbf{u})$, for any $\mathbf{u}$ and $\mathbf{v}$ in V
(iii) $(a\mathbf{u}+b\mathbf{v}, \mathbf{w}) = a(\mathbf{u}, \mathbf{w}) + b(\mathbf{v}, \mathbf{w})$, for any $\mathbf{u}$,$\mathbf{v}$, and $\mathbf{w}$ in V, and any scalars a and b.

The standard inner product on R^n is defined as

$$(\mathbf{u}, \mathbf{v}) = \mathbf{u}^T\mathbf{v}$$

which can be shown that this operation satisfies the above mentioned conditions of an inner product. The inner product $(\mathbf{u}, \mathbf{v}) = \mathbf{u}^T\mathbf{v}$ is also called **dot product**, and $(\mathbf{u}, \mathbf{v})$ is often denoted by $\mathbf{u} \cdot \mathbf{v}$.

Example 1 Given $\mathbf{u} = \begin{bmatrix} 1 \\ 3 \\ 4 \end{bmatrix}$ and $\mathbf{v} = \begin{bmatrix} 2 \\ 0 \\ -5 \end{bmatrix}$. Evaluate (i) $(\mathbf{u}, \mathbf{v})$, (ii) $(\mathbf{v}, \mathbf{u})$, and (iii) $(\mathbf{v}, \mathbf{v})$.

Solution

(i)

$$(\mathbf{u}, \mathbf{v}) = \mathbf{u}^T\mathbf{v}$$

H. Tian, *Linear Algebra*, Synthesis Lectures on Mathematics & Statistics,
https://doi.org/10.1007/978-3-031-84647-2_6

$$= \begin{bmatrix} 1 & 3 & 4 \end{bmatrix} \begin{bmatrix} 2 \\ 0 \\ -5 \end{bmatrix} = (1)(2) + (3)(0) + (4)(-5) = -18$$

(ii)

$$(\mathbf{v}, \mathbf{u}) = \mathbf{v}^T \mathbf{u}$$

$$= \begin{bmatrix} 2 & 0 & -5 \end{bmatrix} \begin{bmatrix} 1 \\ 3 \\ 4 \end{bmatrix} = (2)(1) + (0)(3) + (-5)(4) = -18$$

(iii)

$$(\mathbf{v}, \mathbf{v}) = \mathbf{v}^T \mathbf{v}$$

$$= \begin{bmatrix} 2 & 0 & -5 \end{bmatrix} \begin{bmatrix} 2 \\ 0 \\ -5 \end{bmatrix} = (2)(2) + (0)(0) + (-5)(-5) = 29$$

Following the definition of the standard inner product on R^n, we know that

$$(\mathbf{v}, \mathbf{v}) = \mathbf{v}^T \mathbf{v} = \begin{bmatrix} v_1 & v_2 & \cdots & v_n \end{bmatrix} \begin{bmatrix} v_1 \\ v_2 \\ \vdots \\ v_n \end{bmatrix}$$

$$= v_1^2 + v_2^2 + \cdots v_n^2$$

which is nonnegative. The positive square root of $(\mathbf{v}, \mathbf{v})$ is defined for any $\mathbf{v}$ in R^n. The **length** of a vector $\mathbf{v}$ in R^n is defined by

$$\|\mathbf{v}\| = \sqrt{(\mathbf{v}, \mathbf{v})} = \sqrt{v_1^2 + v_2^2 + \cdots v_n^2}$$

or

$$\|\mathbf{v}\|^2 = (\mathbf{v}, \mathbf{v})$$

We notice that $\|\mathbf{v}\| = 0$ if and only if $\mathbf{v} = \mathbf{0}$. This means that any nonzero vector $\mathbf{v}$ must have a length $\|\mathbf{v}\|$ that is positive.

Example 2 Given $\mathbf{u} = \begin{bmatrix} -1 \\ 3 \end{bmatrix}$ in R^2. Find the length of $\mathbf{u}$.

Solution

$$\|\mathbf{u}\| = \sqrt{(\mathbf{u}, \mathbf{u})} = \sqrt{(-1)^2 + (3)^2} = \sqrt{10}$$

Example 3 Given $\mathbf{v} = \begin{bmatrix} 1 \\ 2 \\ -4 \\ -2 \end{bmatrix}$ in R^4. Find the length of (i) $\mathbf{v}$, (ii) $2\mathbf{v}$, and (iii) $-2\mathbf{v}$.

Solution

(i)

$$\|\mathbf{v}\| = \sqrt{(\mathbf{v}, \mathbf{v})} = \sqrt{(1)^2 + (2)^2 + (-4)^2 + (-2)^2} = \sqrt{1 + 4 + 16 + 4} = \sqrt{25} = 5$$

(ii)

$$\|2\mathbf{v}\| = \sqrt{(2\mathbf{v}, 2\mathbf{v})} = \sqrt{(2)^2 + (4)^2 + (-8)^2 + (-4)^2} = \sqrt{100} = 10$$

(iii)

$$\|-2\mathbf{v}\| = \sqrt{(-2\mathbf{v}, -2\mathbf{v})} = \sqrt{(-2)^2 + (-4)^2 + (8)^2 + (4)^2} = \sqrt{100} = 10$$

For any scalar α and any $\mathbf{v}$ in R^n,

$$\|\alpha\mathbf{v}\| = \sqrt{(\alpha\mathbf{v}, \alpha\mathbf{v})} = \sqrt{(\alpha v_1)^2 + (\alpha v_2)^2 + \cdots + (\alpha v_n)^2} = \sqrt{\alpha^2 \left(v_1^2 + v_2^2 + \cdots v_n^2\right)} = |\alpha| \, \|\mathbf{v}\|$$

So

$$\|\alpha\mathbf{v}\| = |\alpha| \, \|\mathbf{v}\|$$

Using the above formula, for example, for $\alpha = 2$, $\|2\mathbf{v}\| = 2\,\|\mathbf{v}\|$; and when $\alpha = -2$, $\|-2\mathbf{v}\| = 2\,\|\mathbf{v}\|$. The vectors $2\mathbf{v}$ and $-2\mathbf{v}$ have the same length, and they are twice as long as $\mathbf{v}$ in R^n.

A vector with length being 1 is called a **unit vector**. For any nonzero vector $\mathbf{v}$, we can divide it by its length $\|\mathbf{v}\|$ to obtain a unit vector.

Example 4 Consider $\mathbf{v} = \begin{bmatrix} 1 \\ 2 \\ -4 \\ -2 \end{bmatrix}$ in R^4. Find the unit vector with the same direction as $\mathbf{v}$.

Solution *Since* $\|\mathbf{v}\| = 5$, *we divide the given vector* $\mathbf{v}$ *by its length* 5, *or equivalently, multiply* $\mathbf{v}$ *by the scalar* $\frac{1}{5}$, *to obtain a unit vector* $\mathbf{u}$,

$$\mathbf{u} = \left(\frac{1}{\|\mathbf{v}\|}\right)\mathbf{v}$$
$$= \frac{1}{5}\begin{bmatrix} 1 \\ 2 \\ -4 \\ -2 \end{bmatrix}$$
$$= \begin{bmatrix} \frac{1}{5} \\ \frac{2}{5} \\ -\frac{4}{5} \\ -\frac{2}{5} \end{bmatrix}$$

It can be checked that $\|\mathbf{u}\| = 1$.

For any two vectors $\mathbf{x}$ and $\mathbf{y}$ in R^n, the **distance** between them is defined to be $\|\mathbf{x} - \mathbf{y}\|$, i.e., the length of $\mathbf{x} - \mathbf{y}$. Note that $\mathbf{x} - \mathbf{y}$ is just another vector in R^n,

$$\mathbf{x} - \mathbf{y} = \begin{bmatrix} x_1 \\ x_2 \\ \vdots \\ x_n \end{bmatrix} - \begin{bmatrix} y_1 \\ y_2 \\ \vdots \\ y_n \end{bmatrix} = \begin{bmatrix} x_1 - y_1 \\ x_2 - y_2 \\ \vdots \\ x_n - y_n \end{bmatrix}$$

and its length,

$$\|\mathbf{x} - \mathbf{y}\| = \sqrt{(\mathbf{x} - \mathbf{y})^T (\mathbf{x} - \mathbf{y})}$$
$$= \sqrt{(x_1 - y_1)^2 + (x_2 - y_2)^2 + \cdots (x_n - y_n)^2}$$

or

$$\|\mathbf{x} - \mathbf{y}\|^2 = (\mathbf{x} - \mathbf{y})^T (\mathbf{x} - \mathbf{y})$$

Example 5 Given $\mathbf{x} = \begin{bmatrix} 2 \\ 1 \\ 4 \end{bmatrix}$ and $\mathbf{y} = \begin{bmatrix} -1 \\ 0 \\ 2 \end{bmatrix}$ in R^3. Find the distance between $\mathbf{x}$ and $\mathbf{y}$.

Solution *Following the distance formular in R^3,*

$$\begin{aligned}\|\mathbf{x}-\mathbf{y}\| &= \sqrt{(x_1-y_1)^2+(x_2-y_2)^2+(x_3-y_3)^2}\\ &= \sqrt{(2-(-1))^2+(1-0)^2+(4-2)^2}\\ &= \sqrt{3^2+1^2+2^2}=\sqrt{14}\end{aligned}$$

Next, using the properties of the inner product, let's work with $\|\mathbf{x}-\mathbf{y}\|^2$, where $\mathbf{x}$ and $\mathbf{y}$ are arbitrary vectors in R^n.

$$\begin{aligned}\|\mathbf{x}-\mathbf{y}\|^2 &= (\mathbf{x}-\mathbf{y},\mathbf{x}-\mathbf{y})\\ &= (\mathbf{x},\mathbf{x}-\mathbf{y})+(-\mathbf{y},\mathbf{x}-\mathbf{y})\\ &= (\mathbf{x},\mathbf{x}-\mathbf{y})-(\mathbf{y},\mathbf{x}-\mathbf{y})\\ &= (\mathbf{x}-\mathbf{y},\mathbf{x})-(\mathbf{x}-\mathbf{y},\mathbf{y})\\ &= (\mathbf{x},\mathbf{x})-(\mathbf{y},\mathbf{x})-(\mathbf{x},\mathbf{y})+(\mathbf{y},\mathbf{y})\\ &= (\mathbf{x},\mathbf{x})+(\mathbf{y},\mathbf{y})-2\,(\mathbf{x},\mathbf{y})\end{aligned}$$

Therefore, it is true that for any $\mathbf{x}$ and $\mathbf{y}$ in R^n,

$$\|\mathbf{x}-\mathbf{y}\|^2=\|\mathbf{x}\|^2+\|\mathbf{y}\|^2-2\,(\mathbf{x},\mathbf{y})$$

Similarly,

$$\begin{aligned}\|\mathbf{x}+\mathbf{y}\|^2 &= (\mathbf{x}+\mathbf{y},\mathbf{x}+\mathbf{y})\\ &= (\mathbf{x},\mathbf{x}+\mathbf{y})+(\mathbf{y},\mathbf{x}+\mathbf{y})\\ &= (\mathbf{x}+\mathbf{y},\mathbf{x})+(\mathbf{x}+\mathbf{y},\mathbf{y})\\ &= (\mathbf{x},\mathbf{x})+(\mathbf{y},\mathbf{x})+(\mathbf{x},\mathbf{y})+(\mathbf{y},\mathbf{y})\\ &= (\mathbf{x},\mathbf{x})+(\mathbf{y},\mathbf{y})+2\,(\mathbf{x},\mathbf{y})\end{aligned}$$

So, it is also true that for any $\mathbf{x}$ and $\mathbf{y}$ in R^n,

$$\|\mathbf{x}+\mathbf{y}\|^2=\|\mathbf{x}\|^2+\|\mathbf{y}\|^2+2\,(\mathbf{x},\mathbf{y})$$

In either the identity for $\|\mathbf{x}-\mathbf{y}\|^2$ or the identity for $\|\mathbf{x}+\mathbf{y}\|^2$, if $(\mathbf{x},\mathbf{y})=0$, we then have the Pythagorean Theorem.

The two vectors $\mathbf{x}$ and $\mathbf{y}$ in R^n are said to be **orthogonal** to each other if $\mathbf{x}\cdot\mathbf{y}=0$.

Example 6 Given $\mathbf{x}=\begin{bmatrix}5\\1\end{bmatrix}$ and $\mathbf{y}=\begin{bmatrix}-1\\5\end{bmatrix}$ in R^2. Decide whether $\mathbf{x}$ and $\mathbf{y}$ are orthogonal to each other.

Solution *Since*

$$\mathbf{x} \cdot \mathbf{y} = \mathbf{x}^T \mathbf{y} = [5\ 1] \begin{bmatrix} -1 \\ 5 \end{bmatrix} = 0$$

So $\mathbf{x}$ and $\mathbf{y}$ are orthogonal to each other.

Theorem 7 (The Pythagorean Theorem) *The two vectors* $\mathbf{x}$ *and* $\mathbf{y}$ *in* R^n *are orthogonal to each other if and only if*

$$||\mathbf{x} - \mathbf{y}||^2 = ||\mathbf{x}||^2 + ||\mathbf{y}||^2$$

or

$$||\mathbf{x} + \mathbf{y}||^2 = ||\mathbf{x}||^2 + ||\mathbf{y}||^2 .$$

6.1 Exercises

Exercise 8 Let $\mathbf{u} = \begin{bmatrix} 2 \\ -4 \end{bmatrix}$ and $\mathbf{v} = \begin{bmatrix} 6 \\ 3 \end{bmatrix}$. Evaluate (i) $(\mathbf{u}, \mathbf{v})$, (ii) $(5\mathbf{u}, \mathbf{v})$, (iii) $(\mathbf{u}, \mathbf{u})$, (iv) $||\mathbf{u}||$

Exercise 9 Let $\mathbf{x} = \begin{bmatrix} 1 \\ 4 \\ 1 \end{bmatrix}$ and $\mathbf{y} = \begin{bmatrix} 3 \\ 1 \\ 4 \end{bmatrix}$. Evaluate (i) $(\mathbf{x}, \mathbf{y})$, (ii) $(\mathbf{y}, \mathbf{x})$, (iii) $(\mathbf{y}, \mathbf{y})$, (iv) $||\mathbf{y}||$

Exercise 10 Let $\mathbf{u} = \begin{bmatrix} 3 \\ -4 \end{bmatrix}$ in R^2. Find the unit vector with the same direction as $\mathbf{u}$.

Exercise 11 Let $\mathbf{x} = \begin{bmatrix} -1 \\ 1 \\ 2 \end{bmatrix}$ in R^3. Find the unit vector with the same direction as $\mathbf{x}$.

Exercise 12 Let $\mathbf{u} = \begin{bmatrix} 2 \\ -4 \end{bmatrix}$ and $\mathbf{v} = \begin{bmatrix} 6 \\ 3 \end{bmatrix}$. Find the distance between $\mathbf{u}$ and $\mathbf{v}$.

Exercise 13 Let $\mathbf{x} = \begin{bmatrix} 1 \\ 4 \\ 1 \end{bmatrix}$ and $\mathbf{y} = \begin{bmatrix} 3 \\ 1 \\ 4 \end{bmatrix}$. Find the distance between $\mathbf{x}$ and $\mathbf{y}$.

Exercise 14 Decide if $\mathbf{u}$ and $\mathbf{v}$ are orthogonal for each pair of $\mathbf{u}$ and $\mathbf{v}$.

$$\text{(a) } \mathbf{u} = \begin{bmatrix} 2 \\ -4 \end{bmatrix} \text{ and } \mathbf{v} = \begin{bmatrix} 6 \\ 3 \end{bmatrix} \text{ in } R^2$$

$$\text{(b) } \mathbf{u} = \begin{bmatrix} 4 \\ 0 \\ 1 \end{bmatrix} \text{ and } \mathbf{v} = \begin{bmatrix} -3 \\ 3 \\ -1 \end{bmatrix} \text{ in } R^3$$

$$\text{(c) } \mathbf{u} = \begin{bmatrix} 3 \\ -4 \\ 0 \\ 6 \end{bmatrix} \text{ and } \mathbf{v} = \begin{bmatrix} -4 \\ 0 \\ 1 \\ 2 \end{bmatrix} \text{ in } R^4$$

Exercise 15 For $\mathbf{x} = \begin{bmatrix} 3 \\ -7 \end{bmatrix}$ and $\mathbf{y} = \begin{bmatrix} 7 \\ 3 \end{bmatrix}$ in R^2, decide if $\|\mathbf{x} + \mathbf{y}\|^2 = \|\mathbf{x}\|^2 + \|\mathbf{y}\|^2$ is true.

Exercise 16 For $\mathbf{x} = \begin{bmatrix} 2 \\ 1 \\ -1 \end{bmatrix}$ and $\mathbf{y} = \begin{bmatrix} 3 \\ 2 \\ 5 \end{bmatrix}$ in R^3, decide if $\|\mathbf{x} + \mathbf{y}\|^2 = \|\mathbf{x}\|^2 + \|\mathbf{y}\|^2$ is true.

Exercise 17 For $\mathbf{x} = \begin{bmatrix} 2 \\ 1 \\ -1 \end{bmatrix}$ and $\mathbf{y} = \begin{bmatrix} 3 \\ 2 \\ 8 \end{bmatrix}$ in R^3, decide if $\|\mathbf{x} + \mathbf{y}\|^2 = \|\mathbf{x}\|^2 + \|\mathbf{y}\|^2$ is true.

6.2 Orthogonal Sets

Let $\mathbf{v}_1, \mathbf{v}_2, \ldots, \mathbf{v}_k$ be vectors in an inner product space V. If $(\mathbf{v}_i, \mathbf{v}_j) = 0$ for any $i \neq j$, then $\{\mathbf{v}_1, \mathbf{v}_2, \ldots, \mathbf{v}_k\}$ is said to be an **orthogonal set** of vectors.

Example 1 Let $\mathbf{v}_1 = \begin{bmatrix} 1 \\ 1 \\ 0 \\ 0 \end{bmatrix}$, $\mathbf{v}_2 = \begin{bmatrix} 0 \\ 0 \\ 3 \\ 0 \end{bmatrix}$, $\mathbf{v}_3 = \begin{bmatrix} 0 \\ 0 \\ 0 \\ 5 \end{bmatrix}$. The set $\{\mathbf{v}_1, \mathbf{v}_2, \mathbf{v}_3\}$ is an orthogonal set in R^4.

Solution *Since*

$$(\mathbf{v}_1, \mathbf{v}_2) = 0$$
$$(\mathbf{v}_1, \mathbf{v}_3) = 0$$
$$(\mathbf{v}_2, \mathbf{v}_3) = 0$$

So the set is orthogonal.

Theorem 2 *An orthogonal set* $\{\mathbf{v}_1, \mathbf{v}_2, \ldots, \mathbf{v}_k\}$ *of nonzero vectors in an inner product space* V *is a linearly independent set.*

Proof Suppose that the set $\{\mathbf{v}_1, \mathbf{v}_2, \ldots, \mathbf{v}_k\}$ is an orthogonal set of nonzero vectors. To show that the set is linearly independent, let's show that the following equation has trivial solution only.

$$c_1\mathbf{v}_1 + c_2\mathbf{v}_2 + \cdots + c_k\mathbf{v}_k = \mathbf{0}$$

Taking the inner product of $\mathbf{v}_i$, where $i = 1, 2, \ldots, k$, with both sides of the above equation, we obtain

$$c_1(\mathbf{v}_i, \mathbf{v}_1) + c_2(\mathbf{v}_i, \mathbf{v}_2) + \cdots + c_i(\mathbf{v}_i, \mathbf{v}_i) + \cdots + c_k(\mathbf{v}_i, \mathbf{v}_k) = (\mathbf{v}_i, \mathbf{0})$$

Since $\{\mathbf{v}_1, \mathbf{v}_2, \ldots, \mathbf{v}_k\}$ is an orthogonal set, the equation is reduced to,

$$c_i(\mathbf{v}_i, \mathbf{v}_i) = 0$$

Each vector $\mathbf{v}_i$ is a nonzero vector, hence $(\mathbf{v}_i, \mathbf{v}_i) \neq 0$. Thus $c_i = 0$, where $i = 1, 2, \ldots, k$. Therefore, the orthogonal set $\{\mathbf{v}_1, \mathbf{v}_2, \ldots, \mathbf{v}_k\}$ is linearly independent. ■

Corollary 3 *If the set* $\{\mathbf{v}_1, \mathbf{v}_2, \ldots, \mathbf{v}_k\}$ *is an orthogonal set of nonzero vectors in* R^n, *then* $\{\mathbf{v}_1, \mathbf{v}_2, \ldots, \mathbf{v}_k\}$ *is a basis for the subspace Span* $\{\mathbf{v}_1, \mathbf{v}_2, \ldots, \mathbf{v}_k\}$.

Example 4 Let $\mathbf{v}_1 = \begin{bmatrix} 1 \\ -1 \\ 0 \end{bmatrix}$, $\mathbf{v}_2 = \begin{bmatrix} 2 \\ 2 \\ -1 \end{bmatrix}$, $\mathbf{v}_3 = \begin{bmatrix} -1 \\ -1 \\ -4 \end{bmatrix}$. We can verify that the set $\{\mathbf{v}_1, \mathbf{v}_2, \mathbf{v}_3\}$ is an orthogonal set in R^3 because

$$(\mathbf{v}_1, \mathbf{v}_2) = 0$$
$$(\mathbf{v}_1, \mathbf{v}_3) = 0$$
$$(\mathbf{v}_2, \mathbf{v}_3) = 0$$

Then the set $\{\mathbf{v}_1, \mathbf{v}_2, \mathbf{v}_3\}$ is a basis for R^3.

When we use an orthogonal basis for a subspace, we can easily compute the weights in a linear combination.

Theorem 5 *Let the set* $\{\mathbf{v}_1, \mathbf{v}_2, \ldots, \mathbf{v}_k\}$ *be an orthogonal basis for a subspace* V *of* R^n. *Then each* $\mathbf{y}$ *in the subspace* V *can be represented by the basis vectors as follows,*

$$\mathbf{y} = c_1\mathbf{v}_1 + c_2\mathbf{v}_2 + \cdots c_k\mathbf{v}_k$$

with

$$c_i = \frac{(\mathbf{y}, \mathbf{v}_i)}{(\mathbf{v}_i, \mathbf{v}_i)}, \ i = 1, 2, \ldots, k$$

Proof To find c_1, we only need to take the inner product of $\mathbf{v}_1$ with both sides of the equation $\mathbf{y} = c_1\mathbf{v}_1 + c_2\mathbf{v}_2 + \cdots c_k\mathbf{v}_k$. We obtain

$$(\mathbf{v}_1, \mathbf{y}) = c_1 (\mathbf{v}_1, \mathbf{v}_1) + c_2 (\mathbf{v}_1, \mathbf{v}_2) + \cdots + c_i (\mathbf{v}_1, \mathbf{v}_i) + \cdots + c_k (\mathbf{v}_1, \mathbf{v}_k)$$

Since $\{\mathbf{v}_1, \mathbf{v}_2, \ldots, \mathbf{v}_k\}$ is an orthogonal set, the equation is reduced to,

$$(\mathbf{v}_1, \mathbf{y}) = c_1 (\mathbf{v}_1, \mathbf{v}_1)$$

Thus

$$c_1 = \frac{(\mathbf{v}_1, \mathbf{y})}{(\mathbf{v}_1, \mathbf{v}_1)}$$

Similarly, to find c_i, for $i = 2, 3, \ldots, k$, we only need to take the inner product of $\mathbf{v}_i$ with both sides of the equation $\mathbf{y} = c_1\mathbf{v}_1 + c_2\mathbf{v}_2 + \cdots c_k\mathbf{v}_k$. We obtain

$$(\mathbf{v}_i, \mathbf{y}) = c_1 (\mathbf{v}_i, \mathbf{v}_1) + c_2 (\mathbf{v}_i, \mathbf{v}_2) + \cdots + c_i (\mathbf{v}_i, \mathbf{v}_i) + \cdots + c_k (\mathbf{v}_i, \mathbf{v}_k)$$

Since $\{\mathbf{v}_1, \mathbf{v}_2, \ldots, \mathbf{v}_k\}$ is an orthogonal set, the equation is reduced to,

$$(\mathbf{v}_i, \mathbf{y}) = c_i (\mathbf{v}_i, \mathbf{v}_i)$$

So for any $i = 1, 2, \ldots, k$,

$$c_i = \frac{(\mathbf{v}_i, \mathbf{y})}{(\mathbf{v}_i, \mathbf{v}_i)}$$

■

Example 6 We have verified that the set $\{\mathbf{v}_1,\ \mathbf{v}_2,\ \mathbf{v}_3\}$ is an orthogonal basis for R^3, with $\mathbf{v}_1 = \begin{bmatrix} 1 \\ -1 \\ 0 \end{bmatrix}$, $\mathbf{v}_2 = \begin{bmatrix} 2 \\ 2 \\ -1 \end{bmatrix}$, and $\mathbf{v}_3 = \begin{bmatrix} -1 \\ -1 \\ -4 \end{bmatrix}$. Represent $\mathbf{y} = \begin{bmatrix} 5 \\ -3 \\ 1 \end{bmatrix}$ as a linear combination of the vectors in the basis.

Solution *Suppose*

$$\mathbf{y} = c_1\mathbf{v}_1 + c_2\mathbf{v}_2 + c_3\mathbf{v}_3,$$

i.e.,

$$\begin{bmatrix} 5 \\ -3 \\ 1 \end{bmatrix} = c_1 \begin{bmatrix} 1 \\ -1 \\ 0 \end{bmatrix} + c_2 \begin{bmatrix} 2 \\ 2 \\ -1 \end{bmatrix} + c_3 \begin{bmatrix} -1 \\ -1 \\ -4 \end{bmatrix}$$

We have

$$c_1 = \frac{(\mathbf{v}_1, \mathbf{y})}{(\mathbf{v}_1, \mathbf{v}_1)} = \frac{8}{2} = 4$$
$$c_2 = \frac{(\mathbf{v}_2, \mathbf{y})}{(\mathbf{v}_2, \mathbf{v}_2)} = \frac{3}{9} = \frac{1}{3}$$
$$c_3 = \frac{(\mathbf{v}_3, \mathbf{y})}{(\mathbf{v}_3, \mathbf{v}_3)} = \frac{-6}{18} = -\frac{1}{3}$$

So we obtain

$$\begin{bmatrix} 5 \\ -3 \\ 1 \end{bmatrix} = 4\begin{bmatrix} 1 \\ -1 \\ 0 \end{bmatrix} + \frac{1}{3}\begin{bmatrix} 2 \\ 2 \\ -1 \end{bmatrix} - \frac{1}{3}\begin{bmatrix} -1 \\ -1 \\ -4 \end{bmatrix}$$

or

$$\mathbf{y} = 4\mathbf{v}_1 + \frac{1}{3}\mathbf{v}_2 - \frac{1}{3}\mathbf{v}_3.$$

An orthogonal set of unit vectors is called an **orthonormal set**. A set $\{\mathbf{u}_1, \mathbf{u}_2, \ldots, \mathbf{u}_k\}$ is orthonormal if and only if

$$(\mathbf{u}_i, \mathbf{u}_j) = \delta_{ij}$$

where

$$\delta_{ij} = \begin{cases} 1, & \text{if } i = j \\ 0, & \text{if } i \neq j \end{cases}$$

Example 7 The set $\{\mathbf{u}_1, \mathbf{u}_2\}$, where $\mathbf{u}_1 = \begin{bmatrix} \frac{1}{\sqrt{5}} \\ \frac{2}{\sqrt{5}} \end{bmatrix}$ and $\mathbf{u}_2 = \begin{bmatrix} \frac{2}{\sqrt{5}} \\ -\frac{1}{\sqrt{5}} \end{bmatrix}$, is an orthonormal basis for R^2 since

$$(\mathbf{u}_1, \mathbf{u}_2) = 0,\ (\mathbf{u}_1, \mathbf{u}_1) = 1,\ (\mathbf{u}_2, \mathbf{u}_2) = 1$$

From any orthogonal set $\{\mathbf{v}_1, \mathbf{v}_2, \ldots, \mathbf{v}_k\}$, we can obtain an orthonormal set $\{\mathbf{u}_1, \mathbf{u}_2, \ldots, \mathbf{u}_k\}$ by letting

$$\mathbf{u}_i = \left(\frac{1}{\|\mathbf{v}_i\|}\right)\mathbf{v}_i, \quad i = 1, 2, \ldots, k$$

Example 8 From the orthogonal set $\{\mathbf{v}_1, \mathbf{v}_2, \mathbf{v}_3\}$ in R^4, where

$\mathbf{v}_1 = \begin{bmatrix} 1 \\ 1 \\ 0 \\ 0 \end{bmatrix}$, $\mathbf{v}_2 = \begin{bmatrix} 0 \\ 0 \\ 3 \\ 0 \end{bmatrix}$, $\mathbf{v}_3 = \begin{bmatrix} 0 \\ 0 \\ 0 \\ 5 \end{bmatrix}$. We can form an orthonormal set $\{\mathbf{u}_1, \mathbf{u}_2, \mathbf{u}_3\}$

where

$$\mathbf{u}_1 = \left(\frac{1}{\|\mathbf{v}_1\|}\right)\mathbf{v}_1 = \frac{1}{\sqrt{2}}\begin{bmatrix}1\\1\\0\\0\end{bmatrix} = \begin{bmatrix}\frac{1}{\sqrt{2}}\\\frac{1}{\sqrt{2}}\\0\\0\end{bmatrix},$$

$$\mathbf{u}_2 = \left(\frac{1}{\|\mathbf{v}_2\|}\right)\mathbf{v}_2 = \frac{1}{3}\begin{bmatrix}0\\0\\3\\0\end{bmatrix} = \begin{bmatrix}0\\0\\1\\0\end{bmatrix},$$

$$\mathbf{u}_3 = \left(\frac{1}{\|\mathbf{v}_3\|}\right)\mathbf{v}_3 = \frac{1}{5}\begin{bmatrix}0\\0\\0\\5\end{bmatrix} = \begin{bmatrix}0\\0\\0\\1\end{bmatrix},$$

Theorem 9 *Let the set* $\{\mathbf{u}_1, \mathbf{u}_2, \ldots, \mathbf{u}_k\}$ *be an orthonormal basis for a subspace* V *of* R^n. *Then each* $\mathbf{y}$ *in the subspace* V *can be represented by the basis vectors as follows,*

$$\mathbf{y} = c_1\mathbf{u}_1 + c_2\mathbf{u}_2 + \cdots c_k\mathbf{u}_k$$

with

$$c_i = (\mathbf{y}, \mathbf{u}_i), \quad i = 1, 2, \ldots, k.$$

6.2 Exercises

Exercise 10 Let $\mathbf{v}_1 = \begin{bmatrix}1\\1\end{bmatrix}$, $\mathbf{v}_2 = \begin{bmatrix}1\\-1\end{bmatrix}$. Show that the set $S = \{\mathbf{v}_1, \mathbf{v}_2\}$ is an orthogonal set in R^2. Form an orthonormal basis $\{\mathbf{u}_1, \mathbf{u}_2\}$ for R^2.

Exercise 11 Express $\mathbf{y} = \begin{bmatrix}6\\5\end{bmatrix}$ using respectively the orthogonal basis and the orthonormal basis for R^2 given in the previous question.

Exercise 12 Let $\mathbf{v}_1 = \begin{bmatrix}1\\-1\\0\end{bmatrix}$, $\mathbf{v}_2 = \begin{bmatrix}1\\1\\-4\end{bmatrix}$, $\mathbf{v}_3 = \begin{bmatrix}2\\2\\1\end{bmatrix}$. Show that the set $S = \{\mathbf{v}_1, \mathbf{v}_2, \mathbf{v}_3\}$ is an orthogonal basis for R^3. Form an orthonormal basis $\{\mathbf{u}_1, \mathbf{u}_2, \mathbf{u}_3\}$ for R^3.

Exercise 13 Express $\mathbf{y} = \begin{bmatrix} 5 \\ 2 \\ 8 \end{bmatrix}$ using respectively the orthogonal basis and the orthonormal basis for R^3 given in the previous question.

Exercise 14 Let $\mathbf{v}_1 = \begin{bmatrix} 1 \\ 1 \\ 1 \\ 1 \end{bmatrix}$, $\mathbf{v}_2 = \begin{bmatrix} -1 \\ -1 \\ 1 \\ 1 \end{bmatrix}$. Show that the set $S = \{\mathbf{v}_1,\ \mathbf{v}_2\}$ is an orthogonal set in R^4. Form an orthonormal basis $\{\mathbf{u}_1,\ \mathbf{u}_2\}$ for the subspace spanned by S.

Exercise 15 Let $\mathbf{v}_1 = \begin{bmatrix} 1 \\ 0 \\ -1 \\ 0 \end{bmatrix}$, $\mathbf{v}_2 = \begin{bmatrix} 1 \\ 1 \\ 1 \\ 1 \end{bmatrix}$, $\mathbf{v}_3 = \begin{bmatrix} 0 \\ -1 \\ 0 \\ 1 \end{bmatrix}$. Show that the set $S = \{\mathbf{v}_1,\ \mathbf{v}_2,\ \mathbf{v}_3\}$ is an orthogonal set in R^4. Form an orthonormal basis $\{\mathbf{u}_1,\ \mathbf{u}_2,\ \mathbf{u}_3\}$ for the subspace spanned by S.

6.3 Orthogonal Projection

Let $\mathbf{v}$ be a nonzero vector in R^n and we consider a subspace $H = Span\{\mathbf{v}\}$ of R^n. For any vector $\mathbf{y}$ in R^n, we wish to decompose it into the sum of two vectors,

$$\mathbf{y} = \widehat{\mathbf{y}} + \mathbf{p}$$

where $\widehat{\mathbf{y}}$ is in the subspace H, i.e., $\widehat{\mathbf{y}} = \alpha\mathbf{v}$ for some scalar α, and $\mathbf{p}$ is orthogonal to $\mathbf{v}$. We have

$$\mathbf{p} = \mathbf{y} - \widehat{\mathbf{y}} = \mathbf{y} - \alpha\mathbf{v}$$

The vector $\mathbf{p}$ is orthogonal to $\mathbf{v}$ if and only if

$$(\mathbf{y} - \alpha\mathbf{v}) \cdot \mathbf{v} = 0$$

Equivalently,

$$\mathbf{y} \cdot \mathbf{v} - \alpha\mathbf{v} \cdot \mathbf{v} = 0$$

Thus,

$$\alpha = \frac{\mathbf{y} \cdot \mathbf{v}}{\mathbf{v} \cdot \mathbf{v}}$$

Therefore, the vector $\mathbf{y}$ is now written into a sum of two vectors, i.e., $\mathbf{y} = \widehat{\mathbf{y}} + \mathbf{p}$, where

$$\widehat{\mathbf{y}} = \alpha\mathbf{v} = \left(\frac{\mathbf{y} \cdot \mathbf{v}}{\mathbf{v} \cdot \mathbf{v}}\right)\mathbf{v}$$

The vector $\widehat{\mathbf{y}}$ is said to be the **orthogonal projection of y onto v**, or the **orthogonal projection of y onto** H. The $\widehat{\mathbf{y}}$ is also denoted by $proj_{\mathbf{v}}\mathbf{y}$ or $proj_H\mathbf{y}$. The vector $\mathbf{p} = \mathbf{y} - \widehat{\mathbf{y}}$ is said to be the **component of y orthogonal to v**. The distance from $\mathbf{y}$ to the subspace H is $\|\mathbf{y} - \widehat{\mathbf{y}}\|$ or $\|\mathbf{p}\|$.

Example 1 Find the orthogonal projection of $\mathbf{y} = \begin{bmatrix} 2 \\ 8 \end{bmatrix}$ onto $\mathbf{v} = \begin{bmatrix} 1 \\ 1 \end{bmatrix}$. Express $\mathbf{y}$ as the sum of two vectors, one in $Span\{\mathbf{v}\}$ and the other orthogonal to $\mathbf{v}$.

Solution *The orthogonal projection of* $\mathbf{y}$ *onto* $\mathbf{v}$ *is,*

$$\widehat{\mathbf{y}} = \left(\frac{\mathbf{y}\cdot\mathbf{v}}{\mathbf{v}\cdot\mathbf{v}}\right)\mathbf{v} = \left(\frac{10}{2}\right)\begin{bmatrix} 1 \\ 1 \end{bmatrix} = \begin{bmatrix} 5 \\ 5 \end{bmatrix}$$

and the component of $\mathbf{y}$ *orthogonal to* $\mathbf{v}$ *is,*

$$\mathbf{p} = \mathbf{y} - \widehat{\mathbf{y}} = \begin{bmatrix} 2 \\ 8 \end{bmatrix} - \begin{bmatrix} 5 \\ 5 \end{bmatrix} = \begin{bmatrix} -3 \\ 3 \end{bmatrix}$$

So

$$\mathbf{y} = \widehat{\mathbf{y}} + \mathbf{p} = \begin{bmatrix} 5 \\ 5 \end{bmatrix} + \begin{bmatrix} -3 \\ 3 \end{bmatrix}$$

Remark 2 In the above example, the vector $\mathbf{y} = \begin{bmatrix} 2 \\ 8 \end{bmatrix}$ is obviously not in $H = Span\left\{\begin{bmatrix} 1 \\ 1 \end{bmatrix}\right\}$. The $\widehat{\mathbf{y}}$ is like the shadow of $\mathbf{y}$'s in H. The distance from $\mathbf{y}$ to the subspace H is

$$\|\mathbf{y} - \widehat{\mathbf{y}}\| = \|\mathbf{p}\| = 3\sqrt{2}$$

What if $\mathbf{y} = \begin{bmatrix} 8 \\ 8 \end{bmatrix}$? The orthogonal projection of this $\mathbf{y}$ onto $\mathbf{v} = \begin{bmatrix} 1 \\ 1 \end{bmatrix}$ is,

$$\widehat{\mathbf{y}} = \left(\frac{\mathbf{y}\cdot\mathbf{v}}{\mathbf{v}\cdot\mathbf{v}}\right)\mathbf{v} = \left(\frac{16}{2}\right)\begin{bmatrix} 1 \\ 1 \end{bmatrix} = \begin{bmatrix} 8 \\ 8 \end{bmatrix}$$

The $\widehat{\mathbf{y}}$ is the same as $\mathbf{y}$ since the given vector $\mathbf{y}$ is in the subspace H. Then $\mathbf{y} - \widehat{\mathbf{y}} = \begin{bmatrix} 0 \\ 0 \end{bmatrix}$. The distance from $\mathbf{y}$ to the subspace H is

$$\|\mathbf{y} - \widehat{\mathbf{y}}\| = 0.$$

Next, we consider a subspace $H = Span\{\mathbf{v}_1, \mathbf{v}_2\}$, where $\mathbf{v}_1$ and $\mathbf{v}_2$ are orthogonal vectors in R^n. For any vector $\mathbf{y}$ in R^n, we try to decompose it into the sum of two vectors,

$$\mathbf{y} = \widehat{\mathbf{y}} + \mathbf{p}$$

where $\widehat{\mathbf{y}}$ is in the subspace H, i.e., $\widehat{\mathbf{y}} = \alpha_1\mathbf{v}_1 + \alpha_2\mathbf{v}_2$ for some scalars α_1 and α_2, and $\mathbf{p}$ is orthogonal to H. Since

$$\begin{aligned}\mathbf{p} &= \mathbf{y} - \widehat{\mathbf{y}} \\ &= \mathbf{y} - (\alpha_1\mathbf{v}_1 + \alpha_2\mathbf{v}_2) \\ &= \mathbf{y} - \alpha_1\mathbf{v}_1 - \alpha_2\mathbf{v}_2\end{aligned}$$

the vector $\mathbf{p}$ is orthogonal to H if and only if

$$\begin{aligned}(\mathbf{y} - \alpha_1\mathbf{v}_1 - \alpha_2\mathbf{v}_2)\cdot\mathbf{v}_1 &= 0 \\ (\mathbf{y} - \alpha_1\mathbf{v}_1 - \alpha_2\mathbf{v}_2)\cdot\mathbf{v}_2 &= 0\end{aligned}$$

Equivalently,

$$\begin{aligned}\mathbf{y}\cdot\mathbf{v}_1 - \alpha_1\mathbf{v}_1\cdot\mathbf{v}_1 - \alpha_2\mathbf{v}_2\cdot\mathbf{v}_1 &= 0 \\ \mathbf{y}\cdot\mathbf{v}_2 - \alpha_1\mathbf{v}_1\cdot\mathbf{v}_2 - \alpha_2\mathbf{v}_2\cdot\mathbf{v}_2 &= 0\end{aligned}$$

Using the fact that $\mathbf{v}_1$ and $\mathbf{v}_2$ are orthogonal vectors, we have

$$\begin{aligned}\mathbf{y}\cdot\mathbf{v}_1 - \alpha_1\mathbf{v}_1\cdot\mathbf{v}_1 &= 0 \\ \mathbf{y}\cdot\mathbf{v}_2 - \alpha_2\mathbf{v}_2\cdot\mathbf{v}_2 &= 0\end{aligned}$$

So,

$$\alpha_1 = \frac{\mathbf{y}\cdot\mathbf{v}_1}{\mathbf{v}_1\cdot\mathbf{v}_1}, \quad \alpha_2 = \frac{\mathbf{y}\cdot\mathbf{v}_2}{\mathbf{v}_2\cdot\mathbf{v}_2}$$

Therefore, the vector $\mathbf{y}$ is now written into a sum of two vectors, i.e., $\mathbf{y} = \widehat{\mathbf{y}} + \mathbf{p}$, where

$$\begin{aligned}\widehat{\mathbf{y}} &= \alpha_1\mathbf{v}_1 + \alpha_2\mathbf{v}_2 \\ &= \left(\frac{\mathbf{y}\cdot\mathbf{v}_1}{\mathbf{v}_1\cdot\mathbf{v}_1}\right)\mathbf{v}_1 + \left(\frac{\mathbf{y}\cdot\mathbf{v}_2}{\mathbf{v}_2\cdot\mathbf{v}_2}\right)\mathbf{v}_2\end{aligned}$$

is the **orthogonal projection of y onto** H. The vector $\mathbf{p} = \mathbf{y} - \widehat{\mathbf{y}}$ is said to be the **component of y orthogonal to** H. The distance from $\mathbf{y}$ to the subspace H is $\|\mathbf{y} - \widehat{\mathbf{y}}\|$ or $\|\mathbf{p}\|$.

Example 3 Find the orthogonal projection of $\mathbf{y} = \begin{bmatrix} 1 \\ 9 \\ 9 \end{bmatrix}$ onto $H = Span\left\{\begin{bmatrix} 1 \\ 0 \\ -1 \end{bmatrix}, \begin{bmatrix} 1 \\ 0 \\ 1 \end{bmatrix}\right\}$. Express $\mathbf{y}$ as the sum of two vectors, one in H and the other orthogonal to H.

Solution *Note that* $\mathbf{v}_1 = \begin{bmatrix} 1 \\ 0 \\ -1 \end{bmatrix}$ *and* $\mathbf{v}_2 = \begin{bmatrix} 1 \\ 0 \\ 1 \end{bmatrix}$ *are orthogonal. We obtain* $\widehat{\mathbf{y}}$ *as follows,*

$$\begin{aligned}\widehat{\mathbf{y}} &= \left(\frac{\mathbf{y}\cdot\mathbf{v}_1}{\mathbf{v}_1\cdot\mathbf{v}_1}\right)\mathbf{v}_1 + \left(\frac{\mathbf{y}\cdot\mathbf{v}_2}{\mathbf{v}_2\cdot\mathbf{v}_2}\right)\mathbf{v}_2 \\ &= \left(\frac{-8}{2}\right)\begin{bmatrix} 1 \\ 0 \\ -1 \end{bmatrix} + \left(\frac{10}{2}\right)\begin{bmatrix} 1 \\ 0 \\ 1 \end{bmatrix} \\ &= \begin{bmatrix} -4 \\ 0 \\ 4 \end{bmatrix} + \begin{bmatrix} 5 \\ 0 \\ 5 \end{bmatrix} \\ &= \begin{bmatrix} 1 \\ 0 \\ 9 \end{bmatrix}\end{aligned}$$

and the component of $\mathbf{y}$ *orthogonal to* $\mathbf{v}$ *is,*

$$\mathbf{p} = \mathbf{y} - \widehat{\mathbf{y}} = \begin{bmatrix} 1 \\ 9 \\ 9 \end{bmatrix} - \begin{bmatrix} 1 \\ 0 \\ 9 \end{bmatrix} = \begin{bmatrix} 0 \\ 9 \\ 0 \end{bmatrix}$$

So $\mathbf{y}$ *is decomposed as follows,*

$$\mathbf{y} = \widehat{\mathbf{y}} + \mathbf{p} = \begin{bmatrix} 1 \\ 0 \\ 9 \end{bmatrix} + \begin{bmatrix} 0 \\ 9 \\ 0 \end{bmatrix}.$$

Now, we consider a general k-dimensional subspace H in R^n which has an orthogonal basis $\{\mathbf{v}_1, \mathbf{v}_2, \ldots, \mathbf{v}_k\}$. For any vector $\mathbf{y}$ in R^n, we try to decompose it into the sum of two vectors,

$$\mathbf{y} = \widehat{\mathbf{y}} + \mathbf{p}$$

where $\widehat{\mathbf{y}}$ is in the subspace H, i.e., $\widehat{\mathbf{y}} = \alpha_1\mathbf{v}_1 + \alpha_2\mathbf{v}_2 + \cdots + \alpha_k\mathbf{v}_k$ for some scalars α_1, α_2, $\ldots$, and α_k, and $\mathbf{p}$ is orthogonal to H. Since

$$\begin{aligned}\mathbf{p} &= \mathbf{y} - \widehat{\mathbf{y}} \\ &= \mathbf{y} - (\alpha_1\mathbf{v}_1 + \alpha_2\mathbf{v}_2 + \cdots + \alpha_k\mathbf{v}_k) \\ &= \mathbf{y} - \alpha_1\mathbf{v}_1 - \alpha_2\mathbf{v}_2 - \cdots - \alpha_k\mathbf{v}_k\end{aligned}$$

the vector **p** is orthogonal to H if and only if

$$\begin{aligned}(\mathbf{y}-\alpha_1\mathbf{v}_1-\alpha_2\mathbf{v}_2-\cdots-\alpha_k\mathbf{v}_k)\cdot\mathbf{v}_1&=0\\(\mathbf{y}-\alpha_1\mathbf{v}_1-\alpha_2\mathbf{v}_2-\cdots-\alpha_k\mathbf{v}_k)\cdot\mathbf{v}_2&=0\\&\vdots\\(\mathbf{y}-\alpha_1\mathbf{v}_1-\alpha_2\mathbf{v}_2-\cdots-\alpha_k\mathbf{v}_k)\cdot\mathbf{v}_k&=0\end{aligned}$$

Using the fact that $\{\mathbf{v}_1, \mathbf{v}_2, \ldots, \mathbf{v}_k\}$ is an orthogonal basis, we have

$$\begin{aligned}\mathbf{y}\cdot\mathbf{v}_1-\alpha_1\mathbf{v}_1\cdot\mathbf{v}_1&=0\\\mathbf{y}\cdot\mathbf{v}_2-\alpha_2\mathbf{v}_2\cdot\mathbf{v}_2&=0\\&\vdots\\\mathbf{y}\cdot\mathbf{v}_k-\alpha_2\mathbf{v}_k\cdot\mathbf{v}_k&=0\end{aligned}$$

So,

$$\alpha_1=\frac{\mathbf{y}\cdot\mathbf{v}_1}{\mathbf{v}_1\cdot\mathbf{v}_1},\ \alpha_2=\frac{\mathbf{y}\cdot\mathbf{v}_2}{\mathbf{v}_2\cdot\mathbf{v}_2},\ \ldots,\ \alpha_k=\frac{\mathbf{y}\cdot\mathbf{v}_k}{\mathbf{v}_k\cdot\mathbf{v}_k}$$

Therefore, the vector **y** is now written into a sum of two vectors, i.e., $\mathbf{y}=\widehat{\mathbf{y}}+\mathbf{p}$, where

$$\begin{aligned}\widehat{\mathbf{y}}&=\alpha_1\mathbf{v}_1+\alpha_2\mathbf{v}_2+\cdots+\alpha_k\mathbf{v}_k\\&=\left(\frac{\mathbf{y}\cdot\mathbf{v}_1}{\mathbf{v}_1\cdot\mathbf{v}_1}\right)\mathbf{v}_1+\left(\frac{\mathbf{y}\cdot\mathbf{v}_2}{\mathbf{v}_2\cdot\mathbf{v}_2}\right)\mathbf{v}_2+\cdots+\left(\frac{\mathbf{y}\cdot\mathbf{v}_k}{\mathbf{v}_k\cdot\mathbf{v}_k}\right)\mathbf{v}_k\end{aligned}$$

is the **orthogonal projection of y onto** H. The vector $\mathbf{p}=\mathbf{y}-\widehat{\mathbf{y}}$ is said to be the **component of y orthogonal to** H. The distance from **y** to the subspace H is $\|\mathbf{y}-\widehat{\mathbf{y}}\|$ or $\|\mathbf{p}\|$.

6.3 Exercises

Exercise 4 Find the orthogonal projection of $\mathbf{y}=\begin{bmatrix}0\\6\end{bmatrix}$ onto $\mathbf{v}=\begin{bmatrix}-1\\3\end{bmatrix}$. Express **y** as the sum of two vectors, one in *Span* $\{\mathbf{v}\}$ and the other orthogonal to **v**.

Exercise 5 For the previous question, find the distance from $\mathbf{y}=\begin{bmatrix}0\\6\end{bmatrix}$ to the subspace spanned $\mathbf{v}=\begin{bmatrix}-1\\3\end{bmatrix}$.

Exercise 6 Find the orthogonal projection of $\mathbf{y}=\begin{bmatrix}2\\6\\9\end{bmatrix}$ onto $\mathbf{v}=\begin{bmatrix}1\\1\\2\end{bmatrix}$. Express **y** as the sum of two vectors, one in *Span* $\{\mathbf{v}\}$ and the other orthogonal to **v**.

Exercise 7 For the previous question, find the distance from $\mathbf{y} = \begin{bmatrix} 2 \\ 6 \\ 9 \end{bmatrix}$ to the subspace spanned $\mathbf{v} = \begin{bmatrix} 1 \\ 1 \\ 2 \end{bmatrix}$.

Exercise 8 Let H be the subspace in R^3 spanned by the orthogonal basis $\{\mathbf{v}_1, \ \mathbf{v}_2\}$, where $\mathbf{v}_1 = \begin{bmatrix} 1 \\ -1 \\ 0 \end{bmatrix}$, $\mathbf{v}_2 = \begin{bmatrix} 1 \\ 1 \\ -2 \end{bmatrix}$. Express $\mathbf{y} = \begin{bmatrix} 5 \\ 6 \\ 7 \end{bmatrix}$ as the sum of two vectors, one in H and the other orthogonal to H.

Exercise 9 Let H be the subspace in R^4 spanned by the orthogonal basis $\{\mathbf{v}_1, \ \mathbf{v}_2\}$, where $\mathbf{v}_1 = \begin{bmatrix} 1 \\ 1 \\ 1 \\ 1 \end{bmatrix}$, $\mathbf{v}_2 = \begin{bmatrix} -1 \\ -1 \\ 1 \\ 1 \end{bmatrix}$. Express $\mathbf{y} = \begin{bmatrix} 1 \\ 0 \\ 2 \\ 5 \end{bmatrix}$ as the sum of two vectors, one in H and the other orthogonal to H.

Exercise 10 For the previous question, find the distance from $\mathbf{y} = \begin{bmatrix} 1 \\ 0 \\ 2 \\ 5 \end{bmatrix}$ to the subspace H.

Exercise 11 Let H be the subspace in R^4 spanned by the orthogonal basis $\{\mathbf{v}_1, \ \mathbf{v}_2, \ \mathbf{v}_3\}$, where $\mathbf{v}_1 = \begin{bmatrix} 2 \\ 2 \\ 0 \\ 0 \end{bmatrix}$, $\mathbf{v}_2 = \begin{bmatrix} 0 \\ 0 \\ 1 \\ 2 \end{bmatrix}$, $\mathbf{v}_3 = \begin{bmatrix} 0 \\ 0 \\ 2 \\ -1 \end{bmatrix}$. Express $\mathbf{y} = \begin{bmatrix} 5 \\ 2 \\ 8 \\ 6 \end{bmatrix}$ as the sum of two vectors, one in H and the other orthogonal to H.

Exercise 12 For the previous question, find the distance from $\mathbf{y} = \begin{bmatrix} 5 \\ 2 \\ 8 \\ 6 \end{bmatrix}$ to the subspace H.

Exercise 13 Let H be the subspace in R^4 spanned by the orthogonal basis $\{\mathbf{v}_1, \mathbf{v}_2, \mathbf{v}_3\}$, where $\mathbf{v}_1 = \begin{bmatrix} 1 \\ 0 \\ -1 \\ 0 \end{bmatrix}$, $\mathbf{v}_2 = \begin{bmatrix} 1 \\ 1 \\ 1 \\ 1 \end{bmatrix}$, $\mathbf{v}_3 = \begin{bmatrix} 0 \\ -1 \\ 0 \\ 1 \end{bmatrix}$. Express $\mathbf{y} = \begin{bmatrix} 5 \\ 2 \\ 8 \\ 6 \end{bmatrix}$ as the sum of two vectors, one in H and the other orthogonal to H.

Exercise 14 For the previous question, find the distance from $\mathbf{y} = \begin{bmatrix} 5 \\ 2 \\ 8 \\ 6 \end{bmatrix}$ to the subspace H.

GPSR Compliance

The European Union's (EU) General Product Safety Regulation (GPSR) is a set of rules that requires consumer products to be safe and our obligations to ensure this.

If you have any concerns about our products, you can contact us on ProductSafety@springernature.com

In case Publisher is established outside the EU, the EU authorized representative is:

Springer Nature Customer Service Center GmbH
Europaplatz 3
69115 Heidelberg, Germany

Batch number: 08656827

Printed by Printforce, the Netherlands